Geometry and Logic: A Supplement

Second Edition

Margarita Fresquez
Palo Alto College
San Antonio, Texas

Phyllis Arnold
Palo Alto College
San Antonio, Texas

KENDALL/HUNT PUBLISHING COMPANY
4050 Westmark Drive Dubuque, Iowa 52002

ISBN 0-7872-1674-7

Printed in the United States of America
10 9 8 7 6 5 4 3 2 1

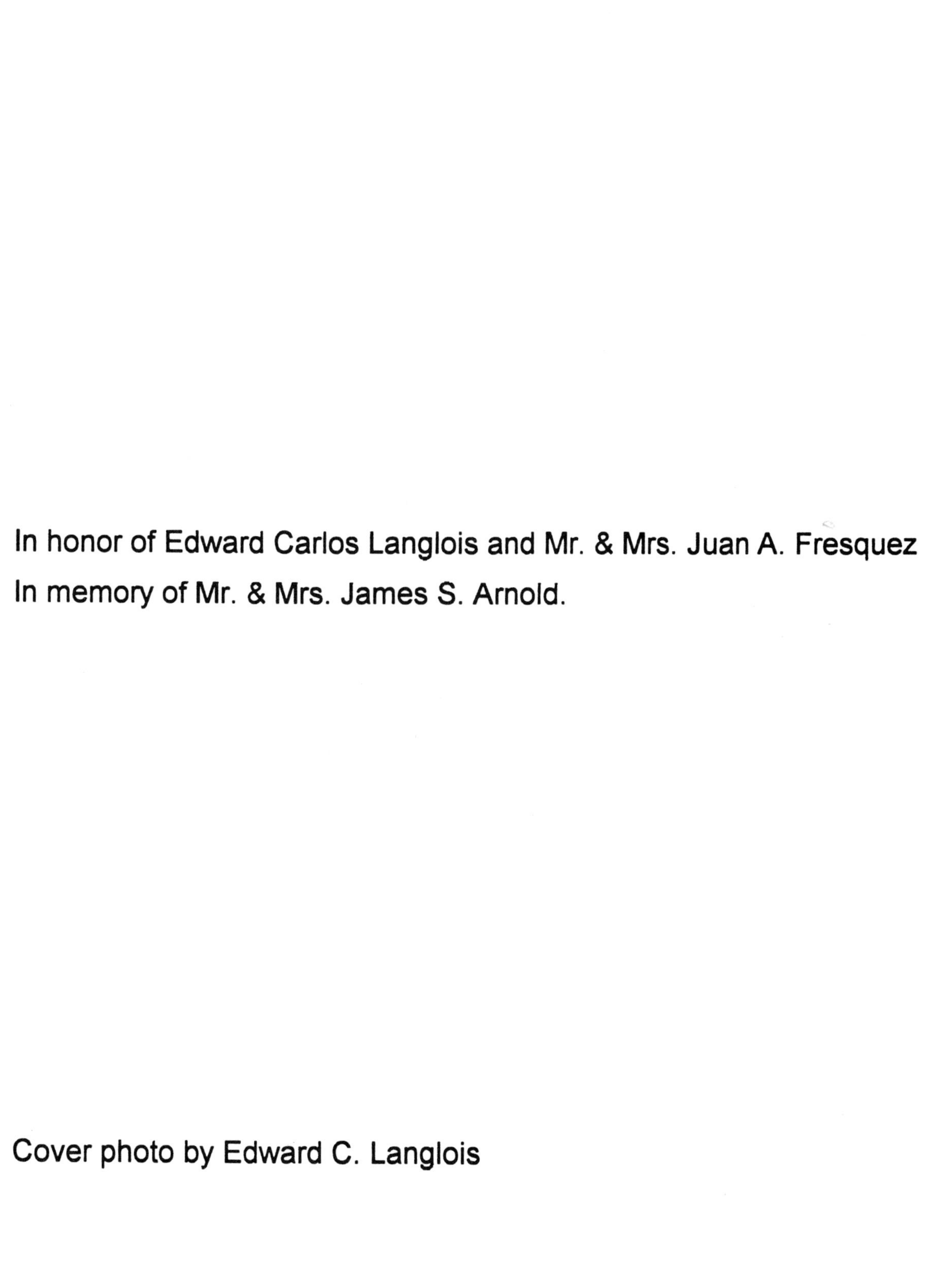

In honor of Edward Carlos Langlois and Mr. & Mrs. Juan A. Fresquez
In memory of Mr. & Mrs. James S. Arnold.

Cover photo by Edward C. Langlois

TABLE OF CONTENTS

Chapter 3 Relationships Between Triangles

PART II - LOGIC

Chapter 4 Logic 1: The Basics

Chapter 5 Logic 2: Reasoning

Chapter 6 Logic 3: Puzzle Problems

Preface

- This book was written to satisfy a need. The pre-algebra and beginning algebra textbooks currently on the market lack sufficient coverage of geometry and logic for student preparation either to pass the TASP (Texas Academic Skills Program) requirement or to be successful in further Mathematics courses. In order to meet these needs, a supplement was necessary. Although it was designed with the specific needs of the Texas student in mind, this book has a substantial treatment of geometry and logic which would be useful for mathematics students in any other state.

- The treatment of the geometry in this book was designed more to solve likely test problems than to serve as a formal axiomatic study of geometry. Showing **where** a geometric situation is likely to occur was considered much more important than any geometic proofs or rigorous treatment of the subject. An attempt has been made to simplify the mystique of logic with careful explanations and examples. This book does not use any symbolic logic, nor does it delve into the more advanced concepts such as truth tables, Boolean algebra, or *exclusive or* which demand a symbolic treatment. The logic contents, however, form a solid base either for further logic as studied in mathematics, computer science, or philosophy

- In both the geometry and logic parts there are solved examples and numerous problems for the student to work after each topic is introduced. These appear with the ⌘ symbol. These "margin exercises" have answers with thorough explanations at the end of their respective chapter.

- Each chapter has one or more exercise sets at the end of the appropriate sections. An attempt has been made to balance the exercise sets so that each type of problem is represented by both an odd-numbered and by an even-numbered problem. All odd-numbered problems have answers in the back of the book with short explanations of steps for the more difficult problems.

ACKNOWLEDGEMENT

- The authors would like to thank Maureen Nowotny for her work in verifying the answer key and Dr. Catalina Fresquez who reviewed this manuscript and provided many valuable suggestions.

PART 1
GEOMETRY

CHAPTER 1 GEOMETRY BASICS

Topics in this chapter:

- ♦ Notation
- ♦ Vertical Angles
- ♦ Parallel Lines
- ♦ Perpendicular Lines
- ♦ Special Pairs of Angles

Answers to ⌘ try-out exercises are given at end of the chapter.
(3b) means that this item or idea is illustrated in **Figure 3, part b.**

1.1 Notation for line, angle, and segment

One of the most important aspects of geometry is the drawing and interpreting of figures to aid in understanding a problem. Such a diagram can aid in both the set-up and solution strategy for the problem. Before we consider an entire geometric figure, we need to learn how to draw and name the parts of it.

POINT: A point has no size, only location. We represent it with a dot. A point is labeled with a capital letter, as in the **point A (1a).**

LINE: A line extends infinitely far out in two opposite directions. We represent the infinitely long line with short straight line with an arrow on either end. One way to name a line is by using a single lower-case letter, as in **line m (1b).** Alternately, any two points on the line may be used to name the line, as in the **line $\overleftrightarrow{CD}$ (1c).**
A line segment is a line drawn between two points, as in the **line segment $\overline{GH}$ (1d).** Please note that the line has a double-headed arrow over the point names, while the segment has just a line over the segment point names.

Figure 1 - lines and points

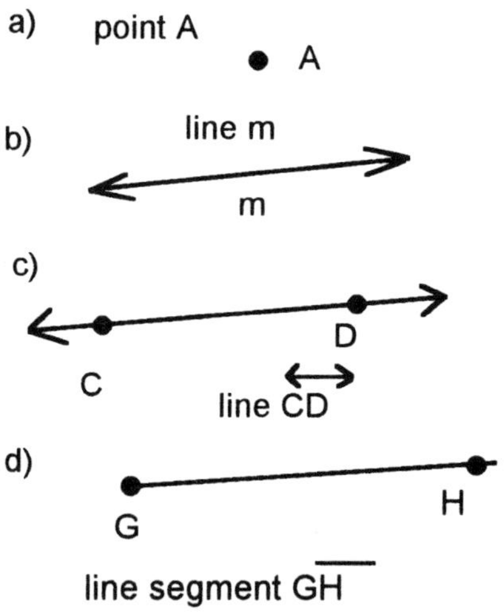

a) point A

A

b) line m

m

c)

C

D

line $\overleftrightarrow{CD}$

d)

G

H

line segment $\overline{GH}$

The **length of the line segment $\overline{GH}$** (between endpoints **G** and **H)** is symbolized by **GH** (no line above the letters.). We do not speak of a line's length, since a line is infinitely long.

Line segments of equal length are called **congruent line segments,** and have a special mark on them in diagrams to show congruence: a "hash" mark (similar to those on football fields) across the line segments. If there are more than two sets of congruent segments, the first would be marked with 1 hash mark, the second with 2 hash marks, etc. In the figure in **(2),** $\overline{DF}$ is marked congruent to $\overline{FE}$, and $\overline{DG}$ is marked congruent to $\overline{GE}$. **Two or more geometrical items are congruent if they are equal in all their measurements.** The symbol for congruence is ≅, and **"segment DF is congruent to segment FE" is symbolized by** $\overline{DF} \cong \overline{FE}$

Figure 2 - congruent line segments

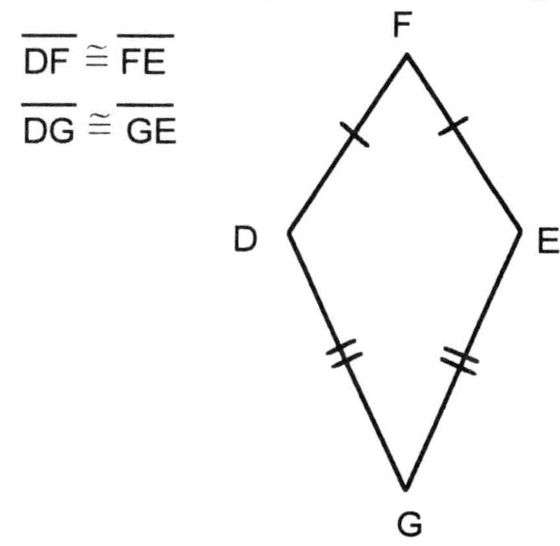

ANGLE: Angles are figures formed by the intersection of two lines. The parts of an angle are the sides and the vertex **(3).** The **vertex** reflects its root meaning "turn", and it is the corner where the angle "turns around". The plural of vertex is vertices.

Figure 3 - the parts of an angle

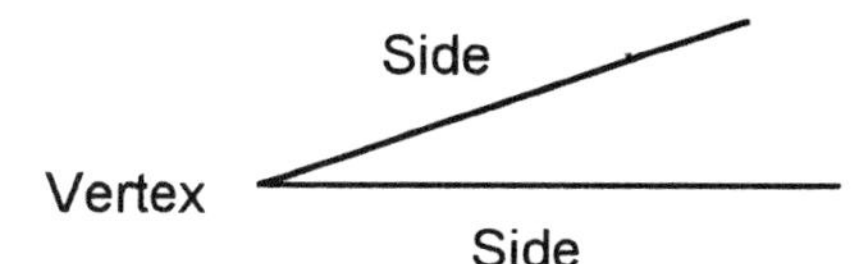

Naming Angles

Angles may be named by one of three methods, illustrated in **Figure 4:**

(1) by using the letter of the vertex point: ∠**A (4a).**

(2) by putting a small number inside the angle: ∠**1 (4a, 4b and 4c)**

(3) by naming three points: the vertex point and one point on each side of the angle. In this case, the vertex point's letter **must** be the middle letter of the three-letter name, but the other two letters can be in either order.

In **(4b),** the letter **A** identifies the point which is the intersection of lines **m** and **n,** and is the vertex of ∠**1.** ∠**1** could also be called ∠**BAC** or ∠**CAB.** It would be inappropriate to refer to ∠**1** as ∠**A** because there are four angles with **A** as their vertex.

In **(4c),** the three-letter name for ∠**1** could be ∠**ABC,** ∠**CBA,** ∠**ABD,** ∠**DBA,** ∠**ABE,** or ∠**EBA.** The important thing is that **B** is the middle letter in each case, indicating that **B** is the vertex angle. The reason that there are several possible names is that the lower side of the angle can be named by any letter along it - **C, D,** or **E.**

Figure 4 - the naming of angles

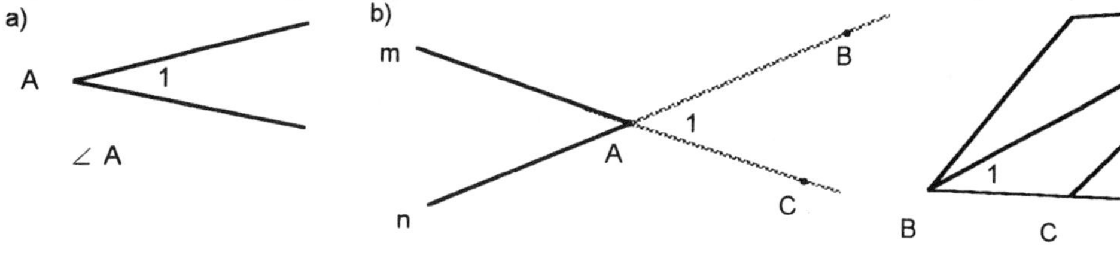

Kinds of Angles

Angles are classified by their angular measurement into one of four categories : acute, right, obtuse, and straight angles:

acute	the measure of the angle is between 0° and 90°
right	the measure of the angle is exactly 90°
obtuse	the measure of the angle is between 90° and 180°
straight	the measure of the angle is exactly 180°

The ° symbol stands for **degrees.** In **Figure 5** are some example angles with their measures. The little square marking in the right angle's vertex is the standard way to indicate a right angle.

Figure 5 - kinds of angles - examples

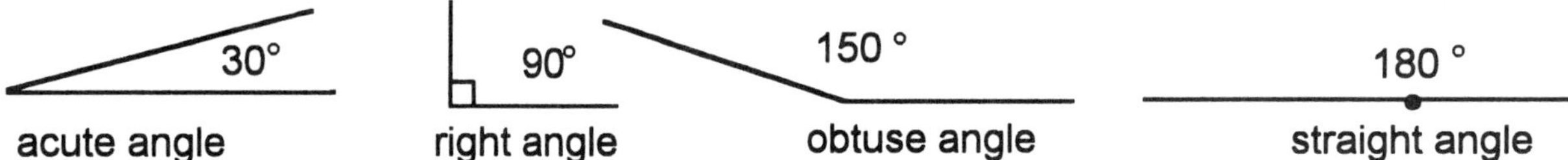

The number of degrees in an angle is called the angle's measure (6a). The **measure of angle A** is represented in symbols by **m∠A.** There are 360° in a whole circle (think of a compass's measurements.) In Geometry, we usually do not consider angles with measure less than 0° or more than 180°. A **protractor** can be used to measure angles. You may find it helpful to get one and draw and measure some angles for practice. **An important thing to keep in mind is that the measure of an angle only reflects the spread of the sides, not their length.** Therefore, **∠A** and **∠B** in **(6b)** have the same measure, even though their sides are of different lengths.

Figure 6 - the measure of an angle

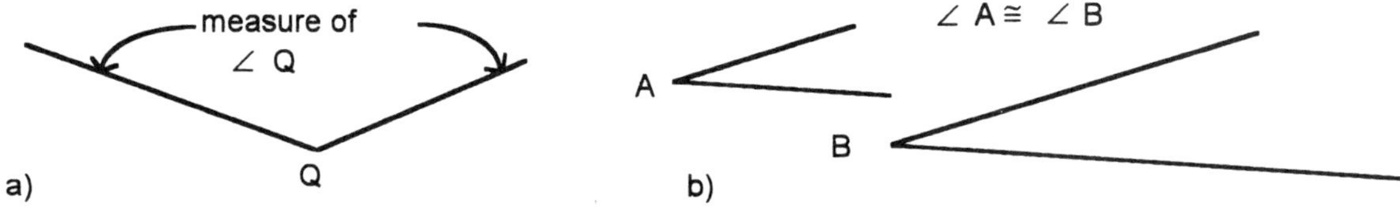

Congruent Angles

Congruent angles are angles with equal measures. For angles A and B above we would write **∠A ≅ ∠B. The standard way to mark congruent angles is by putting arc-marks in the angles.**

If there is more than one set of congruent angles, the first would be marked by 1 arc-mark, the second by 2 arc marks, etc. In the pentagon in **Figure 7, ∠A ≅ ∠B, and ∠D ≅ ∠E**

Figure 7 - congruent angles

C

A B

D E

IMPORTANT NOTE

So far we have talked about two instances of congruence: angles and line segments. To be proper in formal Geometry, when we have two identical items, equal in all possible things that can be measured, we say that these two items are **congruent.** So, equal applies to numbers you get from measurements, and congruence applies to like objects. From this point on, we will deal in informal Geometry, so we will not distinguish between equality and congruence. Where formally we would say, "Congruent angles have equal measures," in this text we will say "Equal angles have the same number of degrees." Instead of "Any triangle which has two congruent angles is an isosceles triangle," we will say "Isosceles triangles have two equal angles."

⌘ **1** In **Drawing 1a,** give the three-letter name for ∠**1,** ∠**2,** and ∠**3**?
In **Drawing 1b,** outline ∠**ECA,** ∠**AEC,** ∠ **ADF,** ∠ **ECG,** and ∠ **EAB** ?

Drawing 1

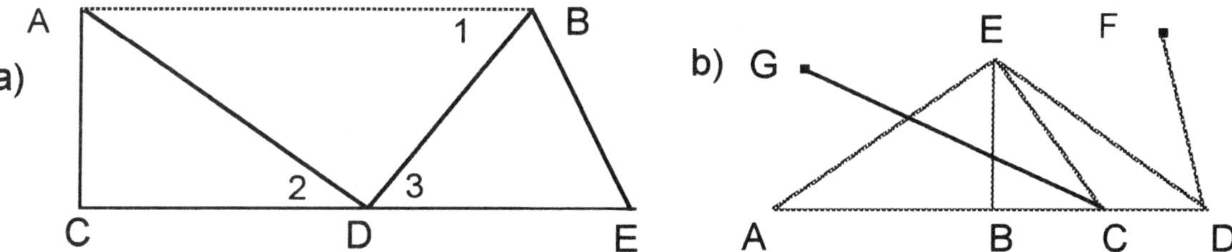

Vertical Angles

When two lines cross, 4 angles are formed. **(8a)** Find the angles which are opposite each other across the intersection of the lines. One pair is 1 and 2, and the other is 3 and 4. Each pair of **vertical angles** consists of two angles which are opposite each other across the **X** made by the lines. The two angles thus formed are connected only by their vertices and are called vertical angles. (This has nothing to do with the usual definition of vertical, or straight up-and-down, but rather it comes from the root vertic- of the word vertices.)

Each pair of intersecting lines will generate two pairs of vertical angles.

Vertical angles are equal.
In **(8b),** by arc-markings, ∠**1** = ∠**2** and ∠**3** = ∠**4.**
Notice that the equal angles are located on either side of the **X.**

Figure 8 - vertical angles

a) 1, 2, 3, 4

b) 1, 2, 3, 4

If the diagrams are complex, such as **(9a),** it can be difficult to determine which angle is vertical to which, and therefore which is equal to ∠**1.** The most dependable way to determine which is the correct vertical angle is to heavily outline the lines bordering ∠**1** ($\overline{AH}$ and $\overline{BH}$), and then extend those lines out past point **H** as in **(9b)** to get ∠**2.** Notice the **X**!!

Figure 9 - figuring out vertical angles

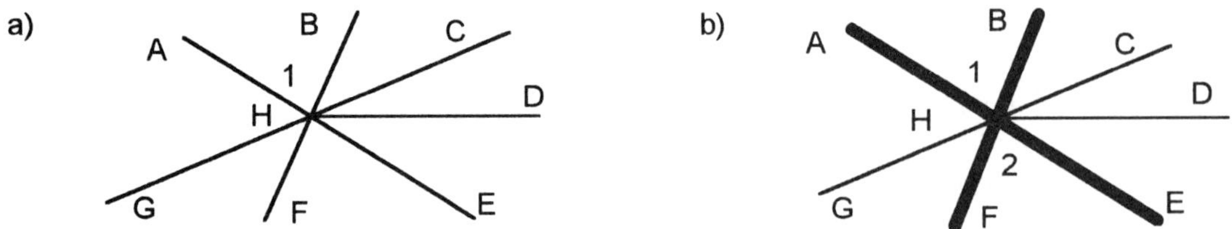

1.3 Parallel lines, alternate angles, and the Z configuration

Parallel lines are lines which never intersect. You can get a good feeling for parallel lines by considering railroad tracks, or the lines containing the right and left sides of a doorway. **Using symbols, if lines m and n are parallel (to each other). we write m || n. In a diagram, we use little chevron-shaped marks to mark a set of parallel lines (10a).** If there is more than one set of parallel lines, the different sets are marked with one, then two, etc. chevrons, as is shown in the parallelogram in **(10b)** with $\overline{AB} \parallel \overline{CD}$ and $\overline{AD} \parallel \overline{BC}$.

Figure 10 - parallel lines

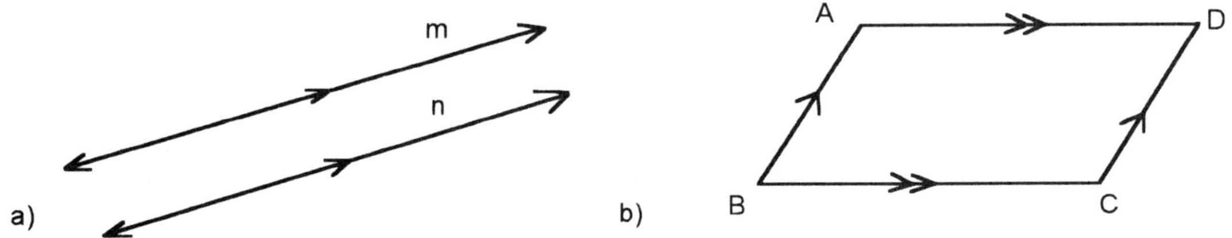

When two parallel lines are cut by a third line, there are two sets of equal angles generated. The official name of the third line is the **transversal.** The two parallel lines and the transversal form a **Z**-shaped figure, with equal angles at the corners of the Z. In **(11a)** below, **m || n,** line **k** is the transversal, and ∠**1** and ∠**2** are the equal angles. The **Z** is heavily outlined. ∠**1** and ∠**2** are in the **alternate interior angle** positions. In **(11b),** we see the other two angles which are commonly named and used as equal angles (∠**2** = ∠**4**). ∠**2** and ∠**4** are in the **corresponding angle** positions, as are ∠**1** and ∠**3.** Take careful note of their location relative to the **Z.** It may be easier to spot the corresponding angles if the top or bottom of the **Z** is extended further **(11b).** There is a third pair of angles commonly named: ∠**2** and ∠**5,** called **same-side interior angles, B.** These angles have a special relationship (supplementarity) which will be considered later. They are not usually equal to each other.

Figure 11 - the Z configuration

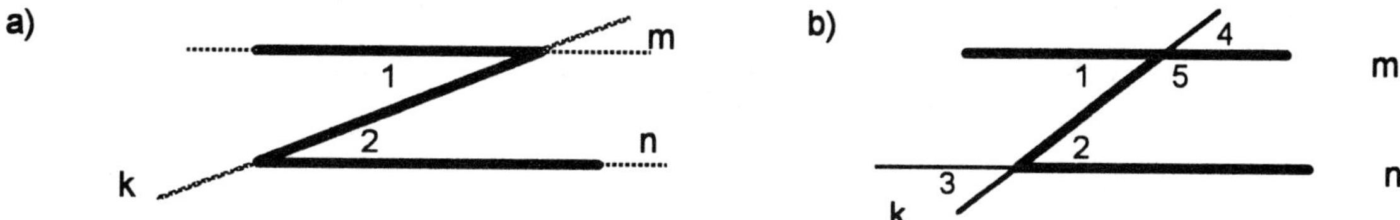

If we draw in all the equal angles, we get two sets of four angles in the pattern shown in **(12a):** one set of equal angles are marked with ●, and the others with ◻. The corresponding arc-marked diagram of this is in **(12b).**

Figure 12 - the Z configuration

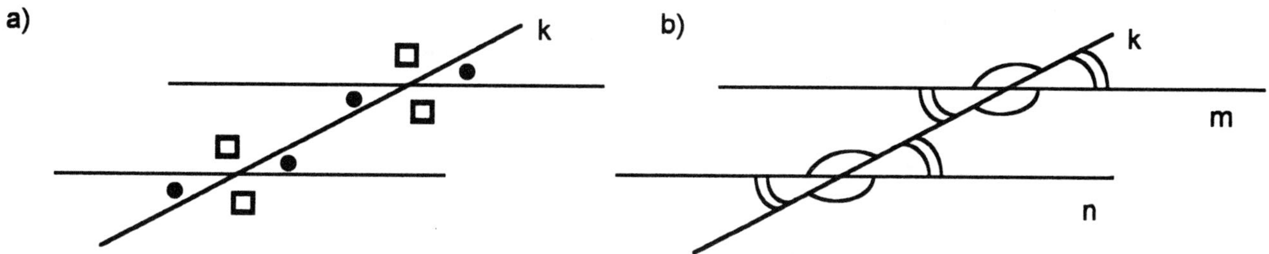

If you ever find it hard to locate the equal angles in a parallel line diagram, try this: extend the parallel lines until it becomes obvious where the **Z** is. The alternate interior angles are located around it. With care, you can also determine where the corresponding angles are.

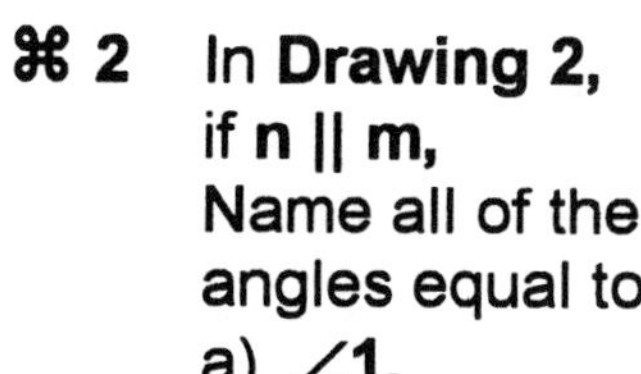

⌘ 2 In **Drawing 2,** if **n || m,** Name all of the angles equal to
a) **∠1,**
b) **∠15,**
c) **∠8?**
Hint: try darkening the **Z.**

Drawing 2

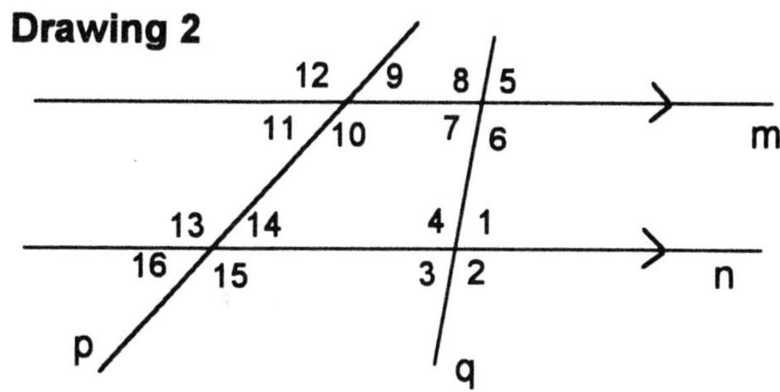

Below are some more situations where it may be difficult to pick out exactly where the parallel lines and the transversals are. In **(13 a and b),** we seek to find **∠1.** (The angles of a triangle add up to 180°.)

Figure 13

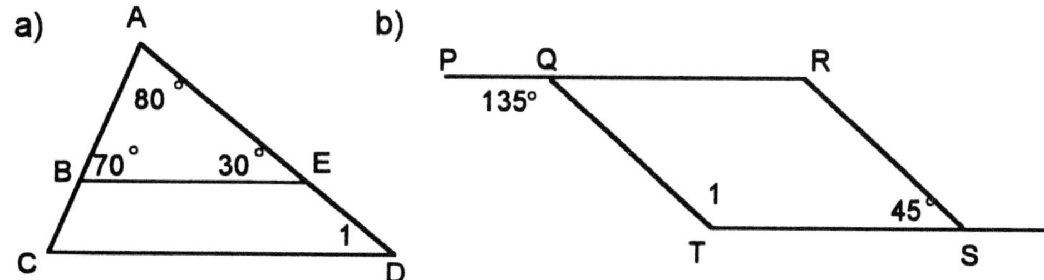

Extending the parallel lines and anything that acts like a transversal can help. In this case, the helpful angle pairs become more apparent. **Figure 13** then turns into **Figure 14.** Then in **(14a),** using corresponding angles, $\angle$**1** = 30°, and in **(14b),** using alternate interior angles, $\angle$**1** = 135°.

Figure 14

a) corresponding angles b) alternate interior angles

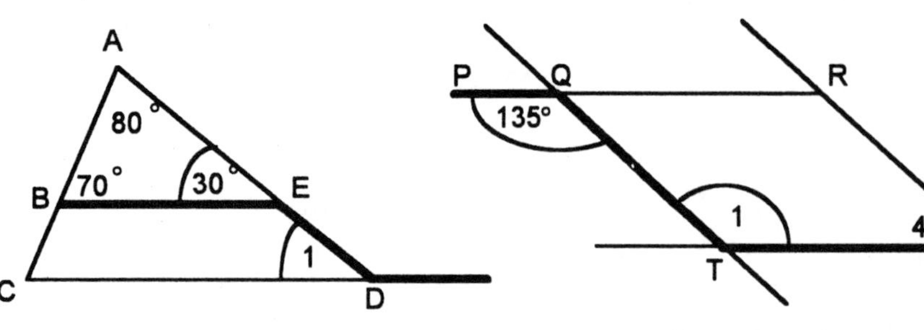

⌘ **3** Consider **Drawing 3,** with **m || n.**

a) Mark the corresponding angle to $\angle$**14**?

b) Mark the alternate interior angle to $\angle$**11**?

c) Think ahead (section 1.5) and guess what $\angle$**1** measures if $\angle$**2** is 55°

d) If $\angle$**10** = 40°, what is $\angle$**13**? What is $\angle$**11**?

Drawing 3

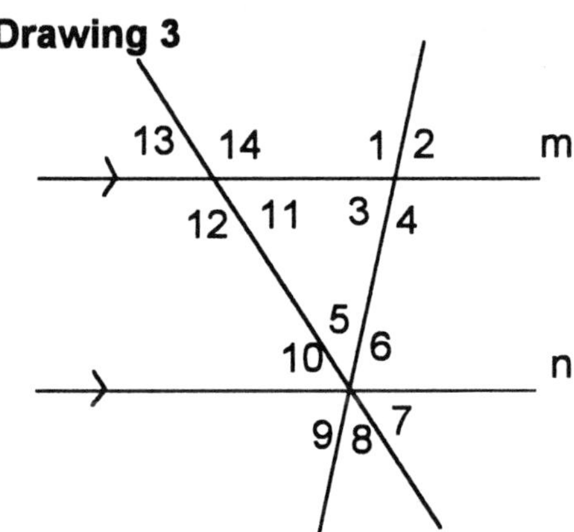

1.4 Perpendicular lines

Perpendicular lines are lines which intersect at right angles to each other (15a). Any time two lines intersect forming one right angle, the other 3 angles formed are also right angles, and all of the right angles are equal to each other **(15b).** The symbol for perpendicular is ⊥. **We write line m is perpendicular to line n with m ⊥ n.** In **(15b), m ⊥ n** implies that $\angle$**1** = $\angle$**2** = $\angle$**3** = $\angle$**4.** In **(15c),** $\angle$**1** = $\angle$**2** implies that **p ⊥ q.**

Figure 15 - perpendicular lines

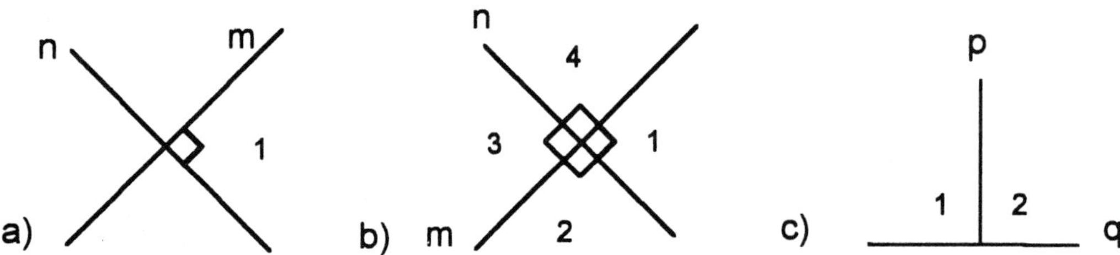

1.5 Special Pairs of angles: Supplementary and complementary angles

There are so many places in which angle pairs adding up to 90° and to 180° occur, that special names have been given to them.

> **Complementary angles are two angles which add up to 90°.**
> **Supplementary angles are two angles which add up to 180°.**

If you have trouble remembering which is which, just remember to associate the "lower" letter, **c,** with the lower number, **90.** In a similar way, you can remember that the "higher" letter, **s,** goes with the higher number, **180.** Here are some of the most common occurrences of complementary angles and supplementary angles.

Figure 16 - complementary and supplemenary angles

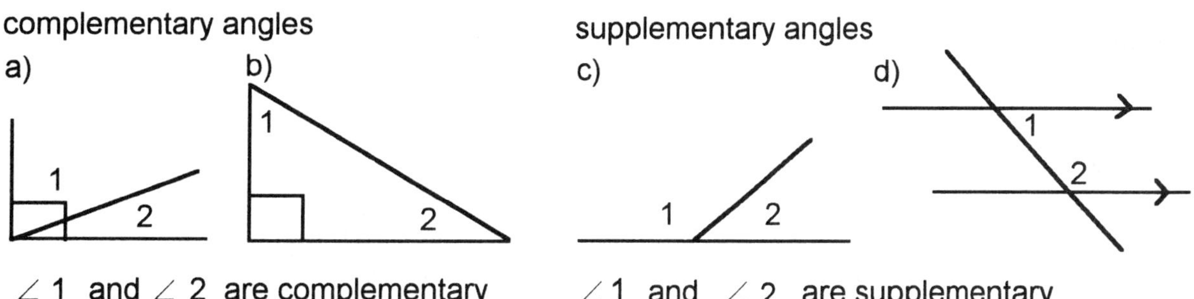

∠ 1 and ∠ 2 are complementary ∠ 1 and ∠ 2 are supplementary

⌘ **4** Can you figure out what x equals in each of the following exercises? Remember that you may have to combine your knowledge about vertical angles and parallel lines with the concepts of supplementarity and complementarity. The extra angles are numbered for easy reference.

Drawing 4

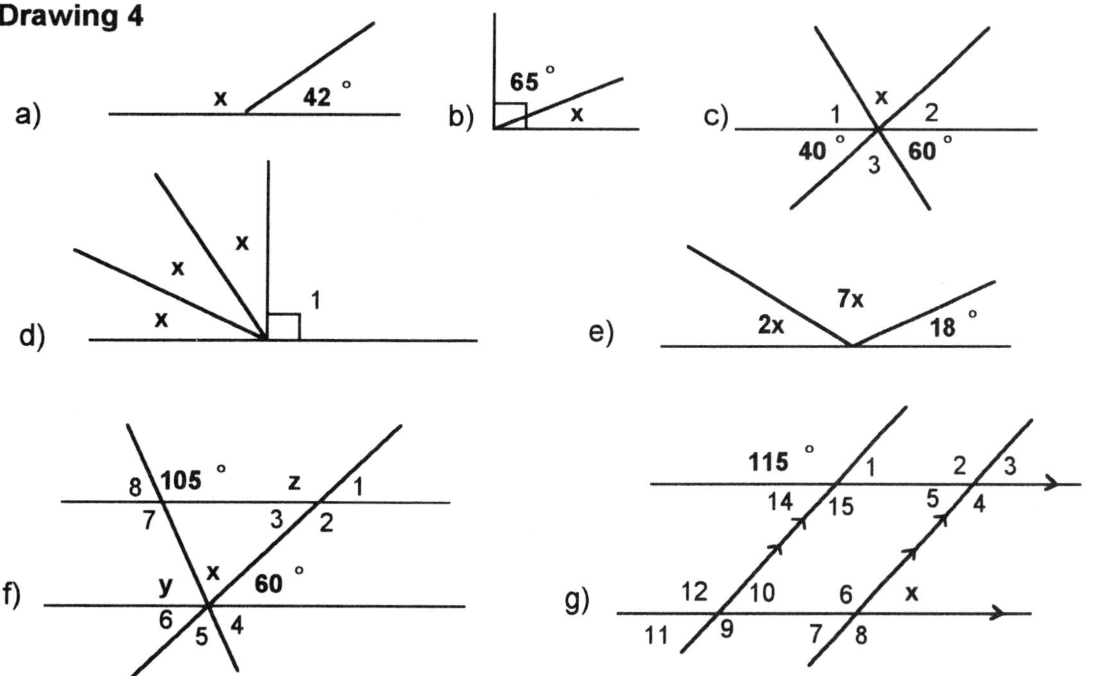

EXERCISES 1.1 - 1.4

1. Match the item names with the items (can have more than one name per item):

items: point, line, line segment, length, angle

item names: ∠ABC, m, $\overline{BC}$, $\overleftrightarrow{AD}$, ∠1, ∠A, k, AB, $\overline{MP}$, ∠M, P

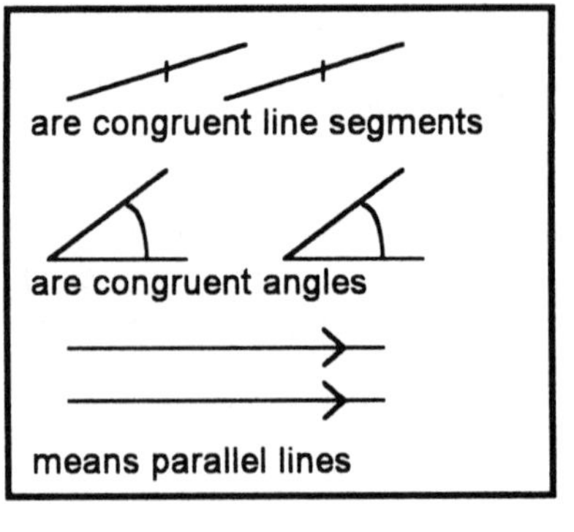

2. Match the item names with the items (can have more than one name per item):

items: point, line, line segment, length, angle

item names: ∠C, AC, n, $\overleftrightarrow{CD}$, P, k, ,$\overline{JK}$ ∠5, DF, L, ∠QFP, $\overline{AF}$

3. Find x:

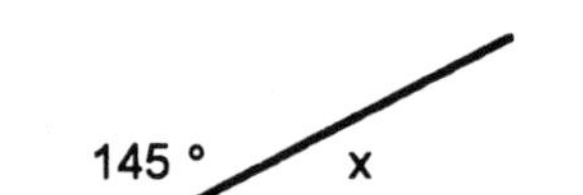

4. Find x:

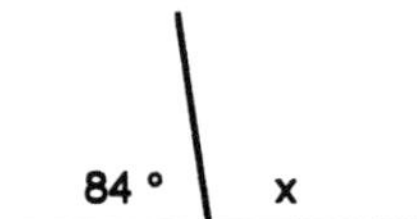

5. ∠1 is the complement of ∠2. If m∠1 = 65°, what is m∠2?

6. ∠3 is the supplement of ∠4. If m∠3 = 122°, what is m∠4?

7. Name the pairs of vertical angles

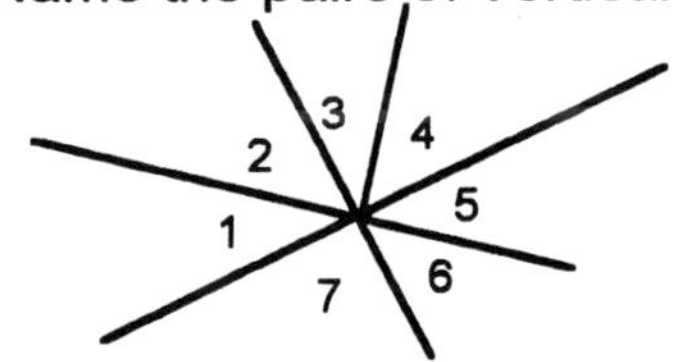

8. Name the pairs of vertical angles

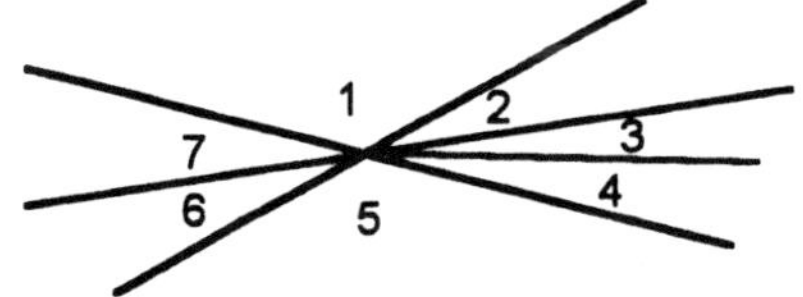

9. Find x, y, z, and w

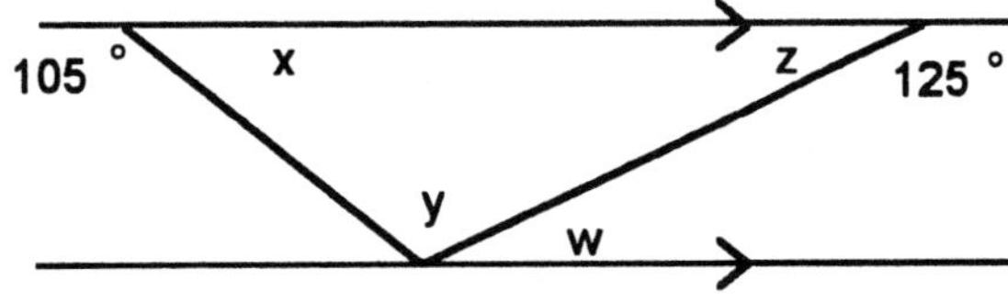

10. Find x, y, z, and w

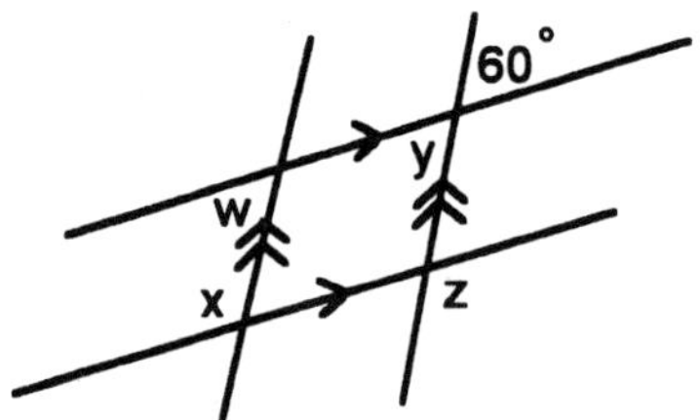

11. Find x, y, z, and w

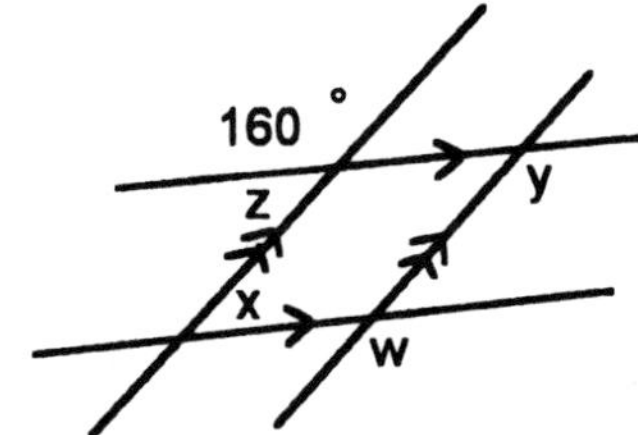

12. Find x, y, and z

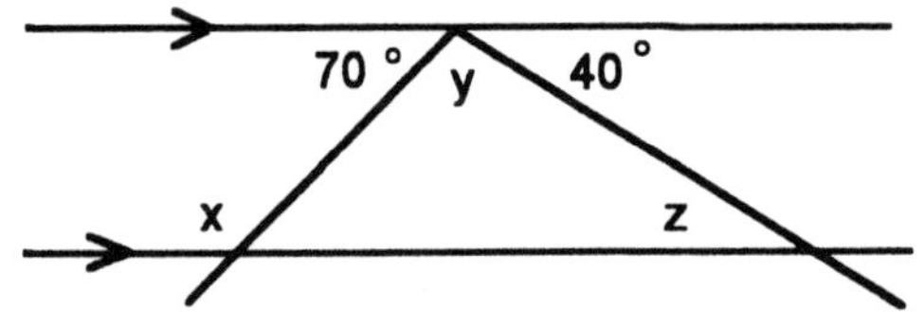

13. Name two pairs each:

a) alternate interior angles
b) corresponding angles
c) same-side interior angles
d) If $\angle 2$ is 118°, what is $\angle 7$?
e) If $\angle 12$ is $\angle 74^{\circ}$, what is $\angle 3$?

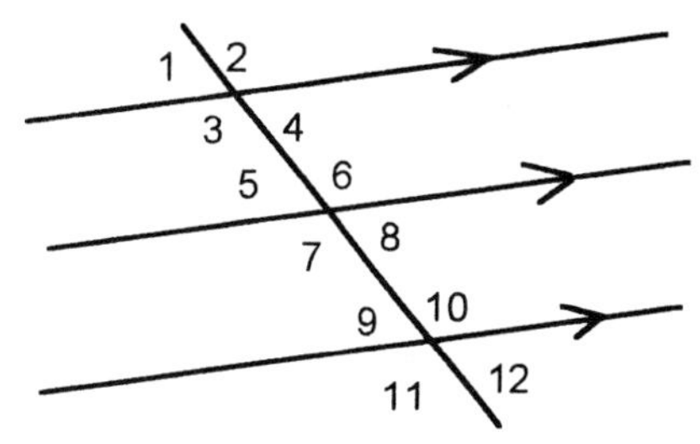

14. Name two pairs each:

a) alternate interior angles
b) corresponding angles
c) same-side interior angles
d) If $\angle 2$ is 115°, what is $\angle 7$?
e) If $\angle 12$ is $\angle 72^{\circ}$, what is $\angle 3$?

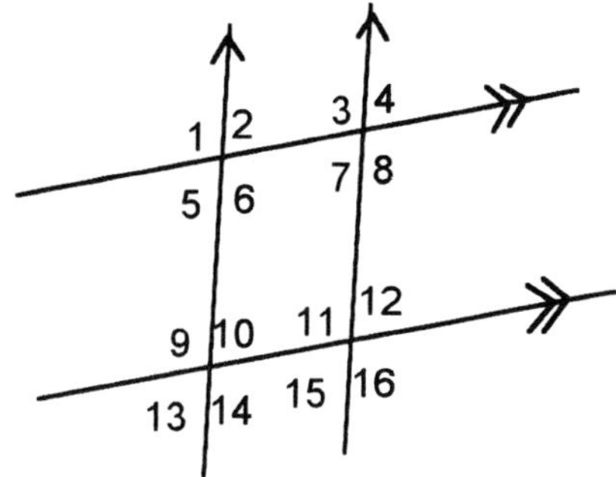

15. Find x.

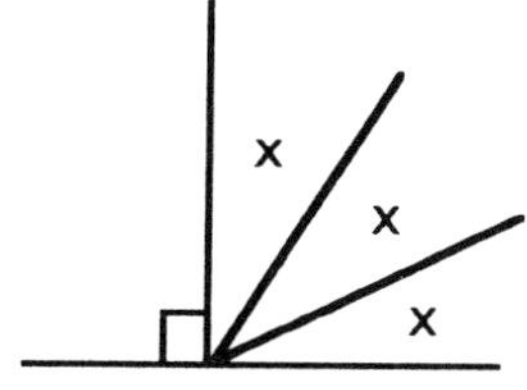

16. Find x.

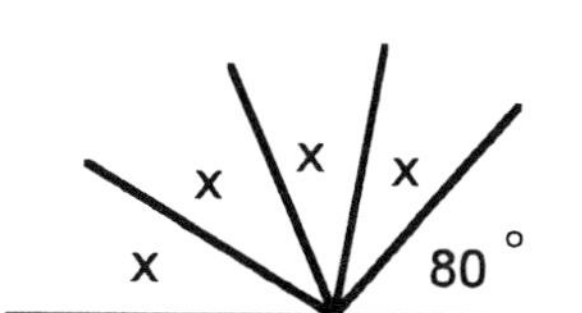

17. Find x

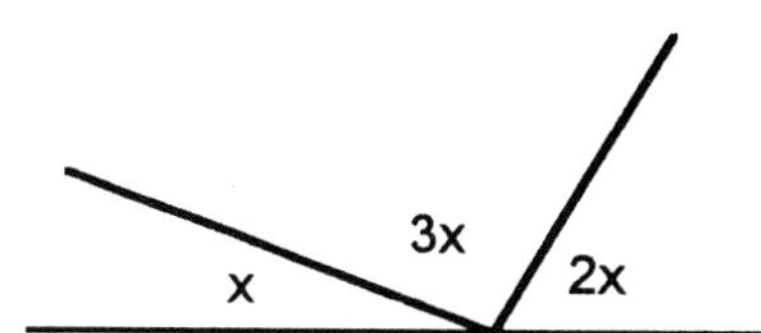

18. Find x.

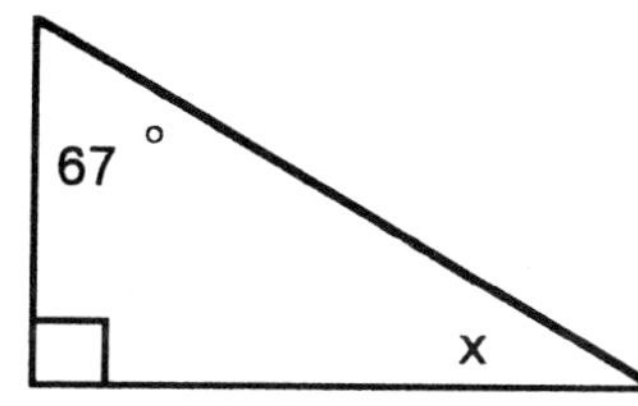

19. Find x.

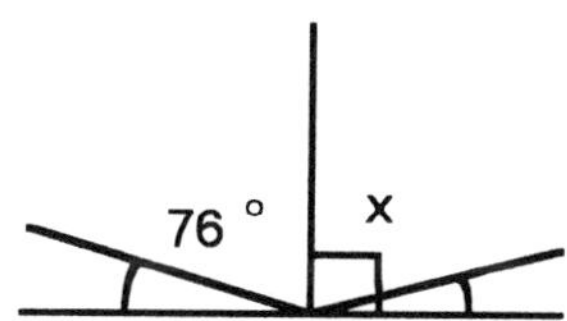

20. Find x.

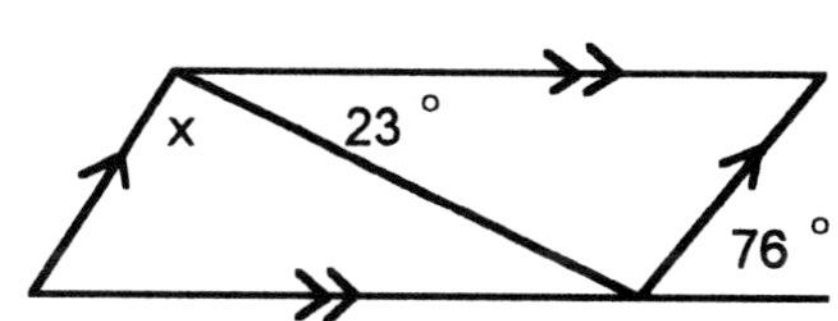

21. Find x

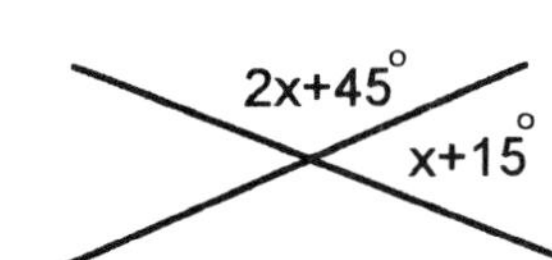

22. Find x

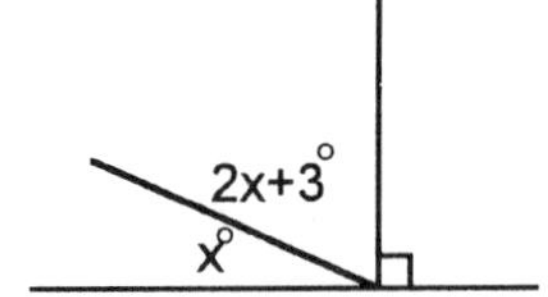

23. Find x

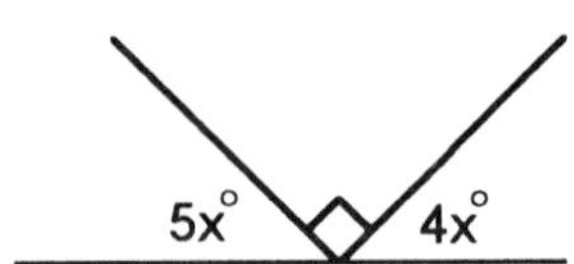

24. Find x

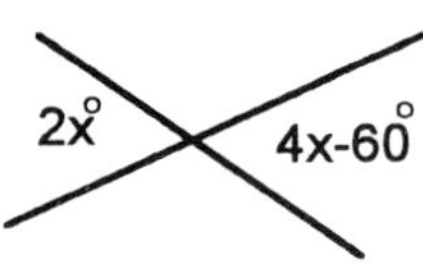

25. Find x

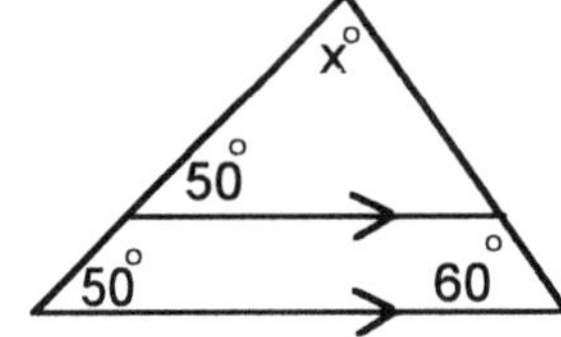

26. Find x

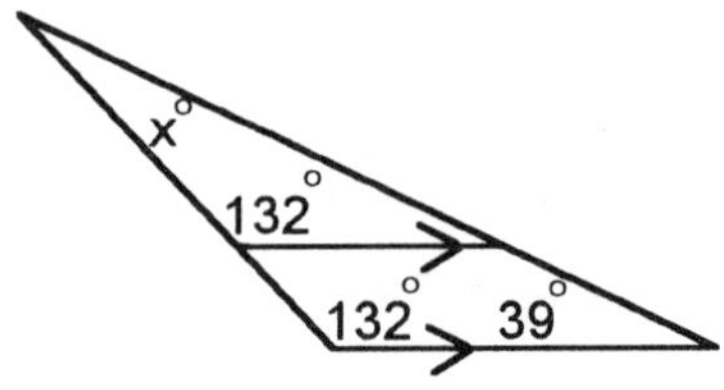

27. Find x and y

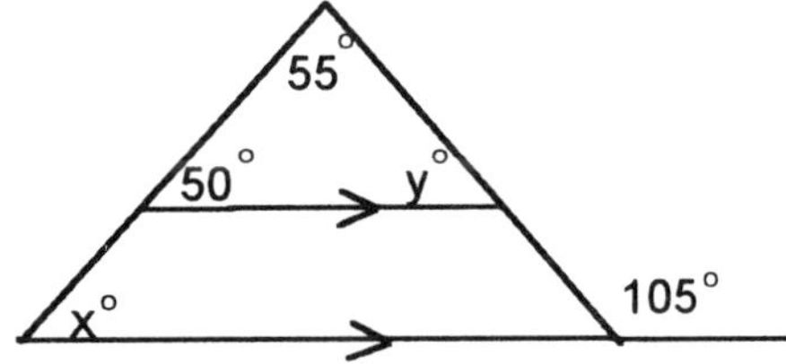

28. Find x and y

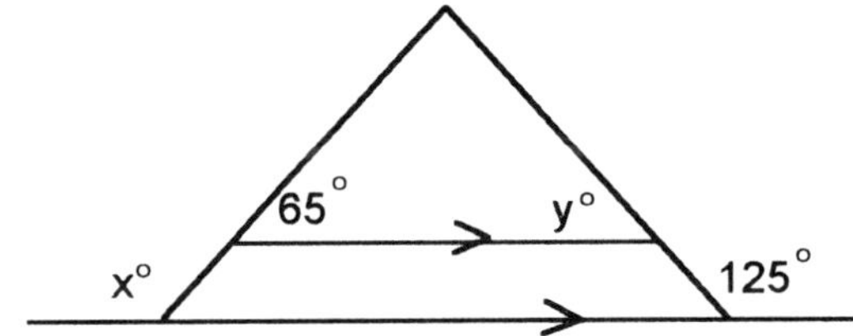

29. An straight angle is made up of two equal angles. How large is each angle?

30. Two complementary angles are congruent. What is the measure of each?

Answers and some reasons to ⌘ problems: Chapter 1

⌘1 For (1a), $\angle 1 = \angle ABD$ or $\angle DBA$, $\angle 2 = \angle ADC$ or $\angle CDA$, $\angle 3 = \angle BDE$ or $\angle EDB$

In (1b) $\angle ECA$ $\angle AEC$ $\angle ADF$ $\angle ECG$ $\angle EAB$

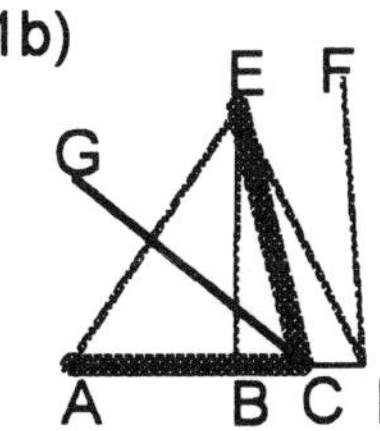

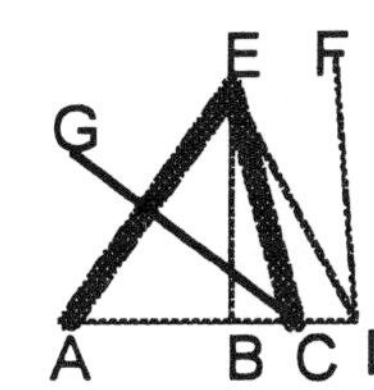

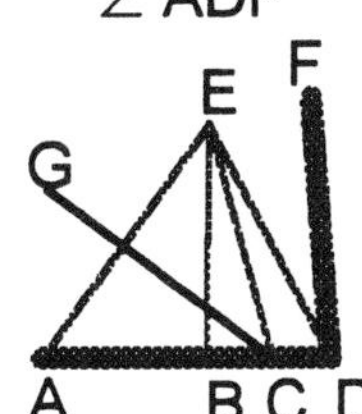

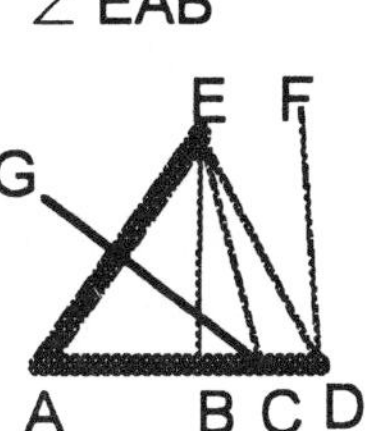

⌘2 The answers are:

a) $\angle 1 = \angle 3, = \angle 5, = \angle 7$ **b)** $\angle 15 = \angle 13, = \angle 10, = \angle 12$

c) $\angle 8 = \angle 6, = \angle 4, = \angle 2$

⌘3 The answers are appropriately arc-marked.

corresponding angles

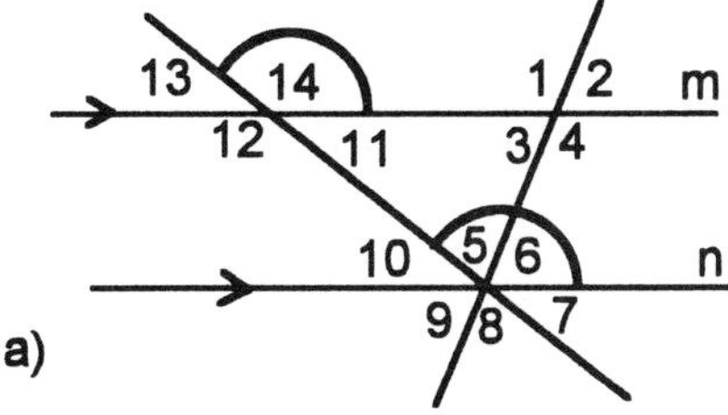

a)

alternate interior angles

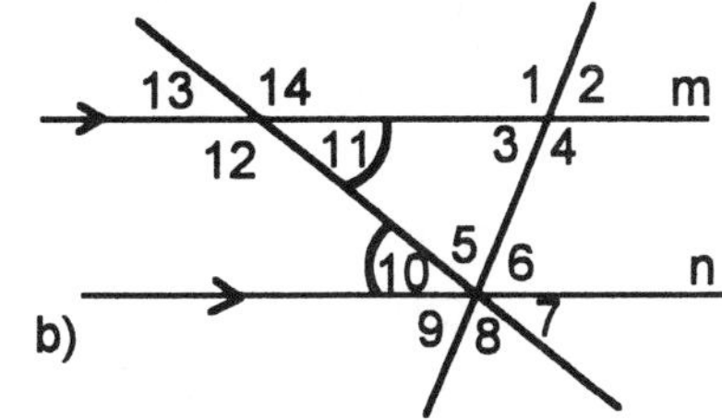

b)

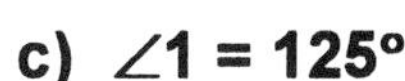

c) $\angle 1 = 125°$

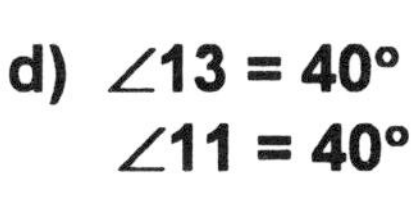

d) $\angle 13 = 40°$
$\angle 11 = 40°$

⌘4 a) $x + 42° = 180°$ (suppl. angles)
$x = 138°$

b) $x + 65° = 90°$ (comp. angles)
$x = 25°$

c) $\angle 3 = x$
$40° + 60° + \angle 3 = 180°$
$\angle 3 = x = 80°$.
Or $\angle 1 = 60°$, $\angle 2 = 40°$
$\angle 1 + x + \angle 2 = 180°$
$60° + x + \angle 4 = 180°$
$x = 80°$.

d) $x + x + x = 90°$
or $(x + x + x + 90° = 180°)$
$3x = 90°$
$x = 30°$

e) $2x + 7x + 18 = 180°$
$9x + 18 = 180°$
$9x = 162°$
$x = 18°$

f) $\angle 8 =$ **$y = 75°$** (corr. angles)

$y + x + 60° = 180°$
$75° + x + 60° = 180°$
$x = 45°$

$z = y + x$ (corr. angle)
$z = 75° + 45°$
$z = 120°$

There are many other correct approaches to this problem.

g) $\angle 2 = 115°$
$\angle 6 = 115°$ (corr. angles)

$x + 115° = 180°$ (suppl angles)
$x = 65°$

There are many other correct approaches to this problem.

CHAPTER 2 POLYGONS

Topics in this chapter:
- ♦ Polygons
- ♦ Triangles
- ♦ Pythagorean Theorem
- ♦ Quadrilaterals

Answers to ⌘ try-out exercises are given at end of the chapter.
(3b) means that this item or idea is illustrated in **Figure 3, part b.**

2.1 Polygons

Polygon is a generic name for a geometric figure which has straight sides. There are two more requirements for a figure to be a polygon:

1) It must be closed, which is a formal way of saying that if you took a pencil and traced around it, you could go all the way around it without ever lifting your pencil or jumping over a gap. Think of a fence around an animal pen. The fence is closed and will keep animals in only if there is no gap in the fence.
2) It must be convex, which means it is not "pinched in" anywhere. Another way to think of it is: if you stretched a rubber band around the figure. If the rubber band hugs the outline all the way around, it is convex.

Figure 1 - polygons

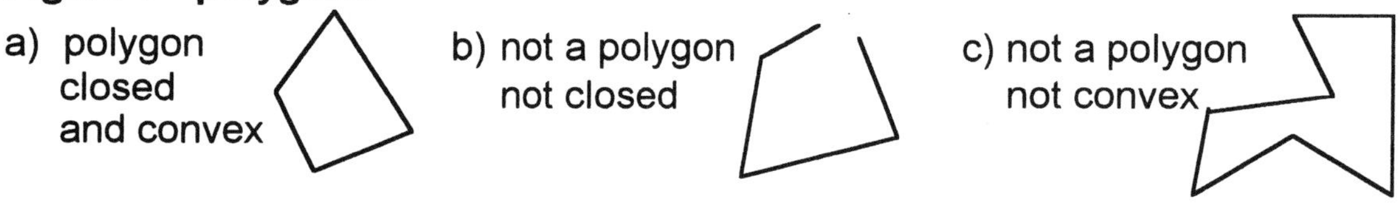

a) polygon closed and convex

b) not a polygon not closed

c) not a polygon not convex

There are many different kinds of polygons. If you think about it a minute, you will realize that the number of sides has to equal the number of angles. The smallest number of sides is 3. "angle" and "agon" mean angle, and "lateral" means side. A polygon has an equal number of sides and angles. The name of a polygon tells you how many sides it has:

triangle	3 sides	octagon	8 sides
quadrilateral	4 sides	nonagon	9 sides
pentagon	5 sides	decagon	10 sides
hexagon	6 sides	n-gon	n sides
heptagon	7 sides	polygon	many angles

Figure 2 shows some of the most common polygons.

If the figure is star-shaped; it cannot be convex. Such a figure is usually given a name ending in -angle, such as the pentangle in **(2f).**

Figure 2 - polygons

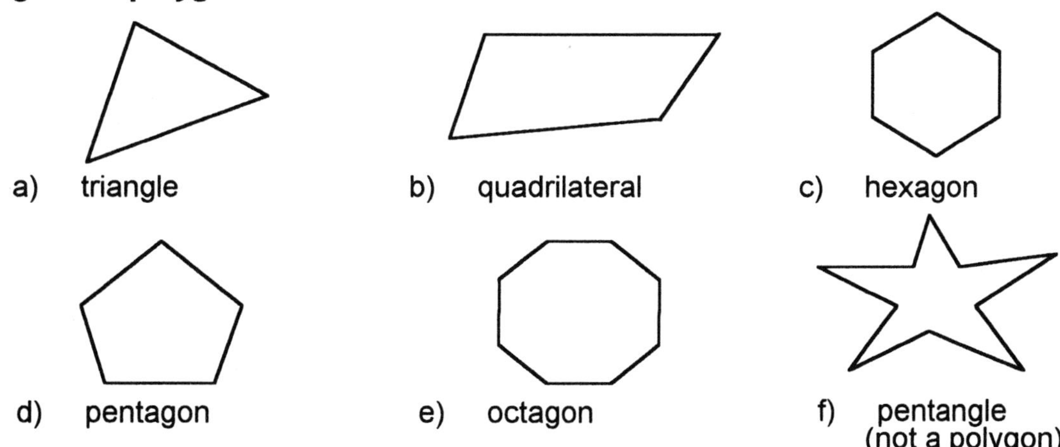

Instances of the triangle in everyday life are not hard to think of: the YIELD sign, architectural support designs, love triangles, religious symbols, etc. Quadrilaterals are the basis of most construction today. Also, most desks, this book, and a sheet of paper are quadrilaterals. The most famous pentagon is The Pentagon in Washington, DC., but there are numerous examples of 5-sided patterns in flowers and invertebrate sea life. Hexagons show up in honeycombs, crystals, architecture, art, and religious symbols. The most common occurrence of the octagon is the STOP sign.

A polygon is called a regular polygon if all of its angles and sides are equal. A regular triangle is called an equilateral triangle, and a regular quadrilateral is called a square. **(2c), (2d),** and **(2e)** are as close to perfect regular polygons as could readily be drawn.

⌘ 1 See how many of the polygonal shapes you can find in the world around you.

2.2 Triangles

Triangles have three angles, and three sides. **The angles of a triangle add up to 180°,** regardless of what the triangle looks like.

Triangles are frequently referred to by the nature of their angles: acute, right, and obtuse.

> The **acute triangle** is made up of 3 acute angles.
> The **right triangle** is made up of one right angle, and a pair of acute angles.
> The **obtuse triangle** is made up of one obtuse angle and 2 acute angles.

Triangles can also be classified by what kind of sides they have.

A **scalene triangle** has no two sides which are the same length.
An **isosceles triangle** has at least two sides of equal length. It also has two equal angles opposite the equal sides. This is a very special triangle which we will cover in detail later.
The **equilateral triangle** has all three sides equal, and is also equiangular (all three angles equal to each other.)

Figure 3 - triangles named by their sides and angles

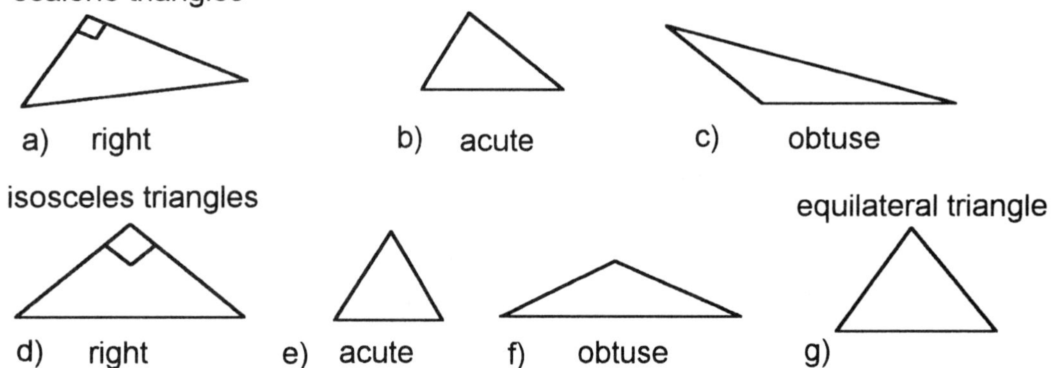

Both isosceles and scalene triangles can come in the right, acute, or obtuse variety. There is only one kind of equilateral triangle: acute.

The special anatomy of the isosceles triangle:

Figure 4 - isosceles triangle

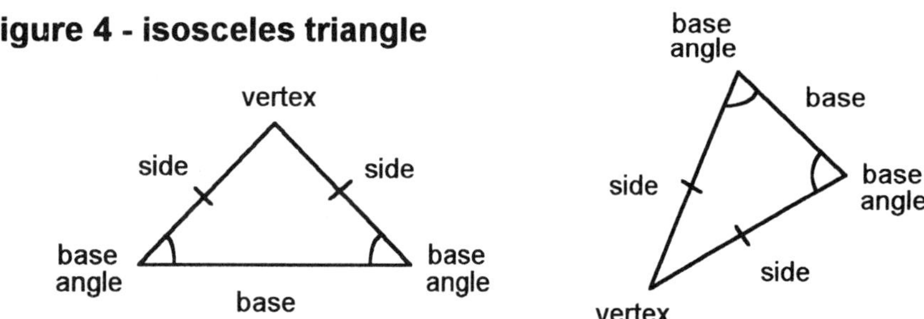

The orientation of the isosceles triangle is immaterial; the odd-angle-out is always called the **vertex,** and the side opposite it (not touching the vertex in any way) is called the **base.** The two equal angles (on either end of the base) are called the **base angles,** and the other two sides are called the **sides,** or **legs,** and are equal to each other.

The special anatomy of the right triangle:
The sides of a triangle which form the right angle are called **legs** (or **sides**), and the side opposite the right angle is called the **hypotenuse.** The hypotenuse is always the longest side of the right triangle. The two acute angles of the right triangle are always complementary.

Figure 5 - right triangle

In **2.3,** we will see that there is a special relationship between the sides of any right triangle, given by the Pythagorean Theorem.

⌘ 2 Find the indicated letter(s) for both of the figures in **Drawing 1.**

Drawing 1

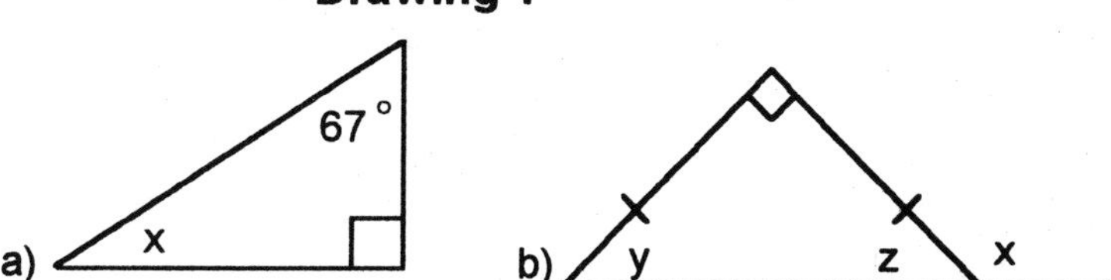

Two consequences of the fact that the sum of the angles of any triangle is 180°:

1) **third angle principle: in two triangles, if two angles of one of the triangles are equal to two corresponding angles of the other triangle, the third angles must also be equal.**

2) **exterior angle principle: the exterior angle of a triangle is equal to the sum of the two non-adjacent interior angles**

Figure 6 - the two principles illustrated

a) third angle principle

if you start out with

1
65°
20°

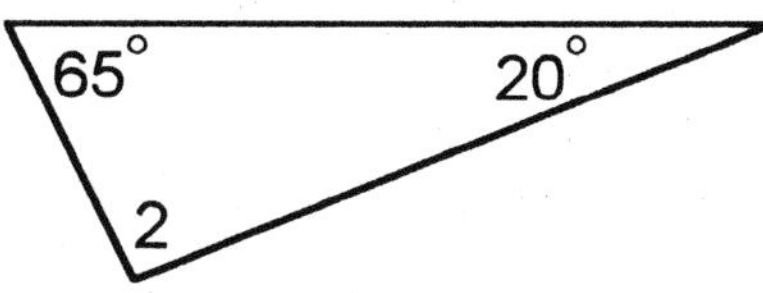

you can assert that $\angle$**1** ≅ $\angle$**2,** and each angle measures 95°.

b) exterior angle principle

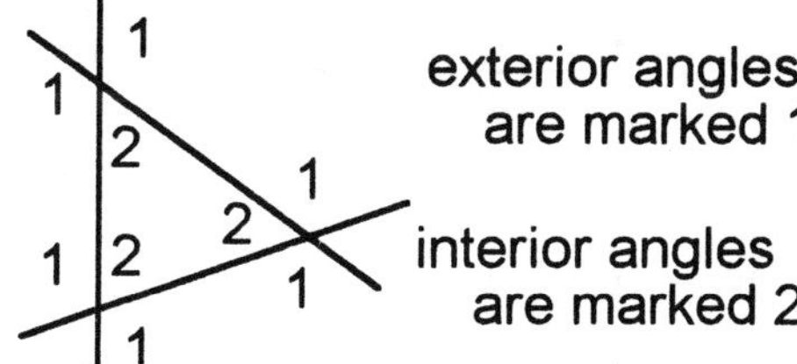

if you start with this:

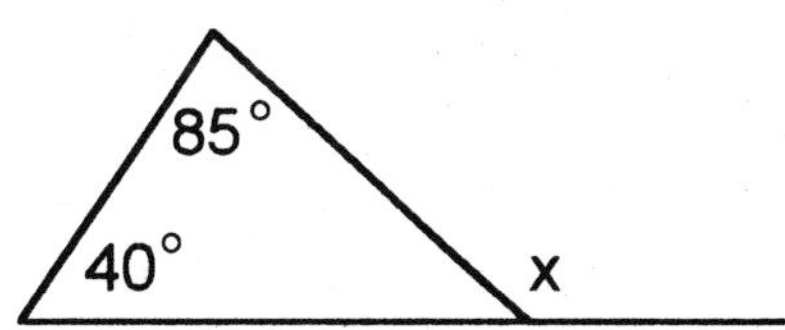

you can assert that x = 85° + 40° = 125°

⌘ 3 Find the indicated letter(s) for each of the figures in **Drawing 2.**

Drawing 2

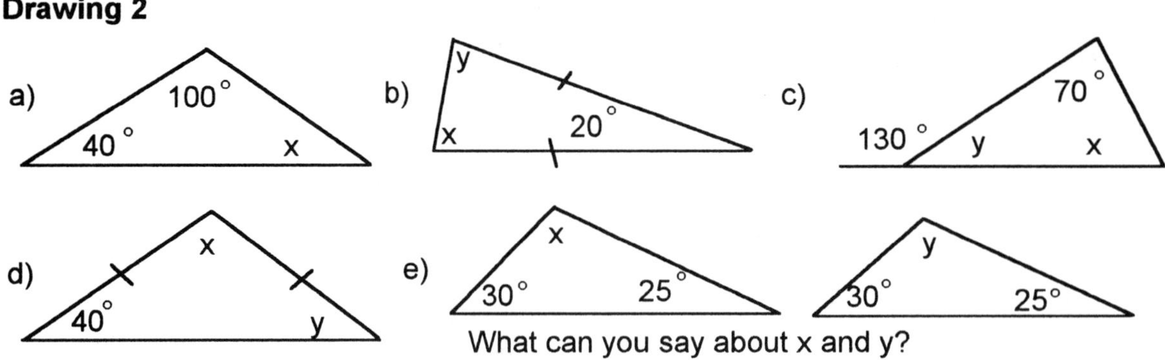

What can you say about x and y?

In any triangle, **a perpendicular line drawn from any vertex to the opposite side is called an altitude.** When you are computing areas of triangles, the height in the area formula ($A = \frac{1}{2}bh$) is the length of the altitude, and the base in the area formula is the side to which it is drawn. In a right triangle, the two legs can serve as base and height.

Figure 7 - altitudes

for an acute triangle:

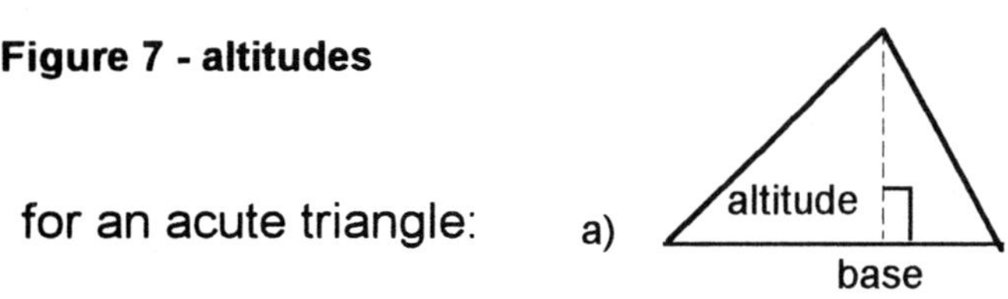

Figure 7 - altitudes

for an acute triangle: the two possibilities for a right triangle:

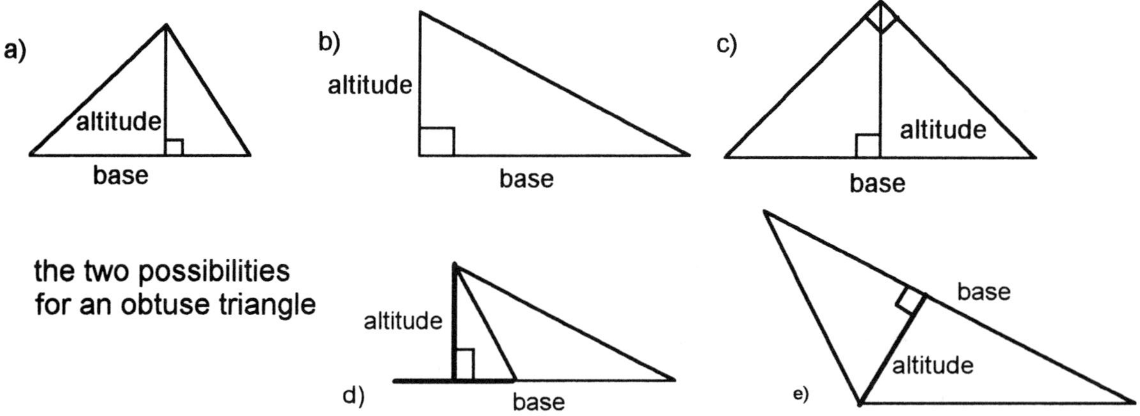

the two possibilities for an obtuse triangle

Dividing anything into two equal parts is called **bisecting. In an isosceles triangle, the altitude to the base bisects the triangle into two congruent triangles.**

There are several consequences of this. The most important one is that the altitude to the base of an isosceles triangle divides the base into two equal parts. The other consequences are most easily remembered this way:

If you have a triangle with a line in it from a vertex to the side opposite that vertex, any two of the following facts imply the other two (8):

1) **the triangle is isosceles**
2) **the line is an altitude to the base**
3) **the line bisects the side to which it was drawn**
4) **the line bisects the angle at the vertex from which it was drawn.**

Figure 8 - altitude to the base of an isosceles triangle

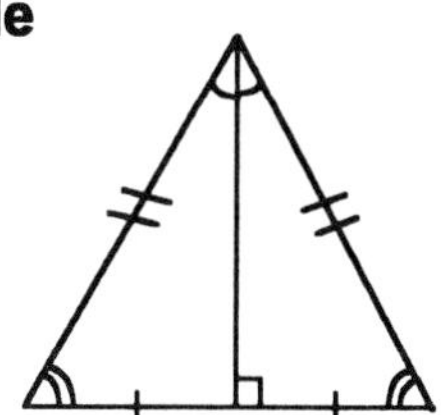

EXERCISES 2.1 & 2.2

1. Find x:

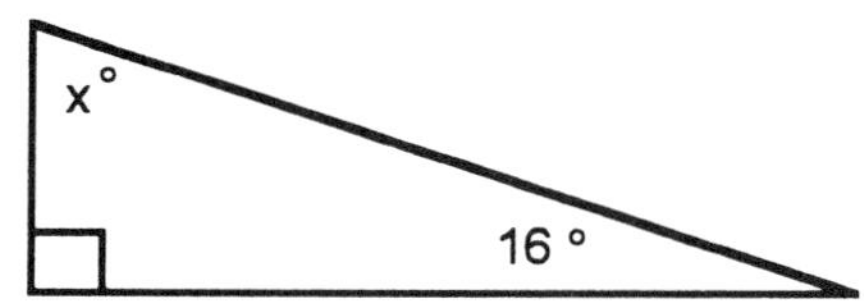

2. Find x:

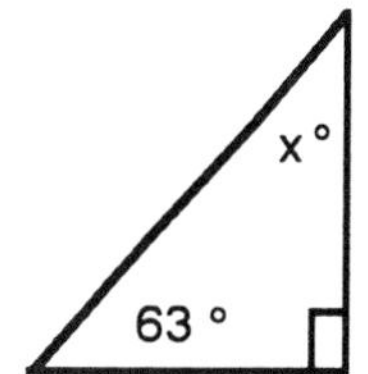

3. Find x:

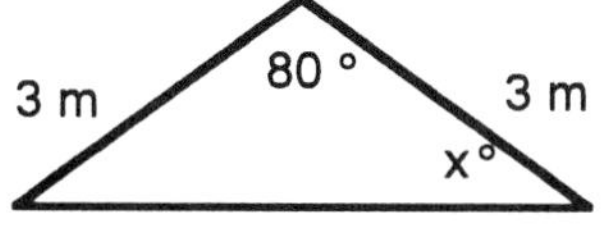

4. Find x:

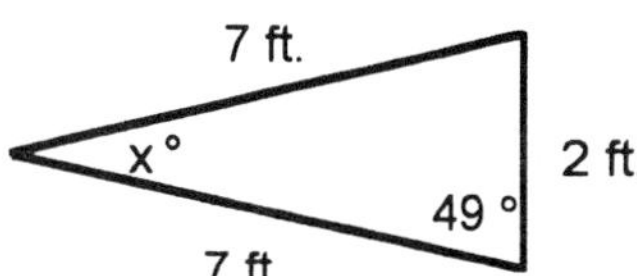

5. Find x:

Δ ABC is a right triangle

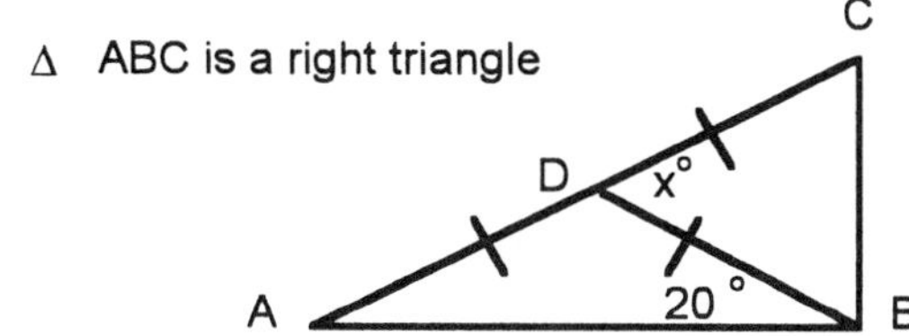

6. Find x:

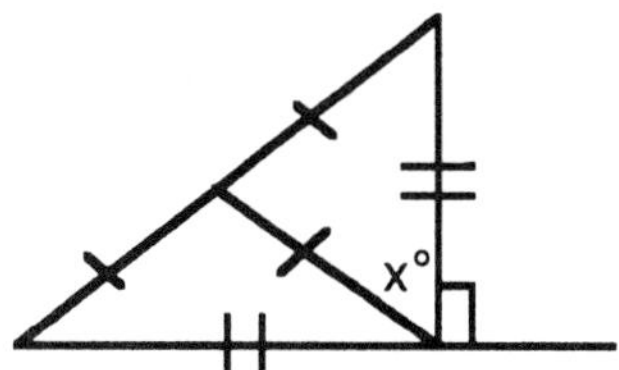

7. Find x:

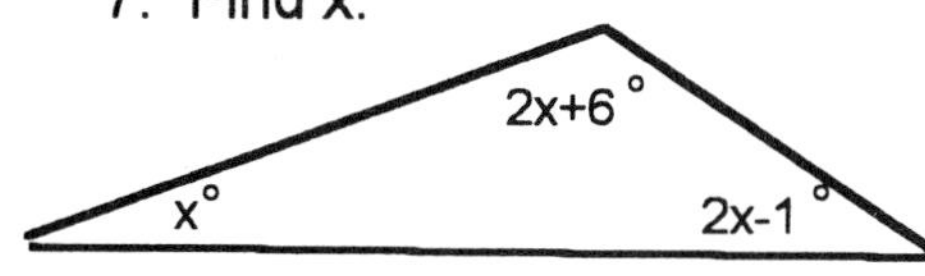

8. Find x:

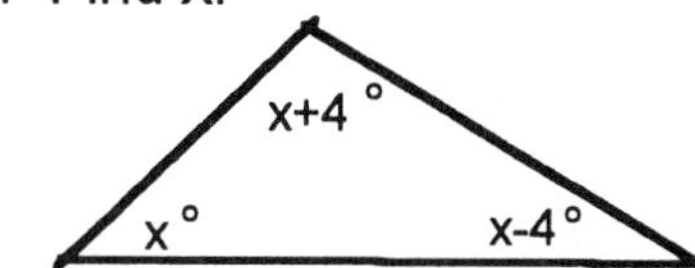

9. Find x

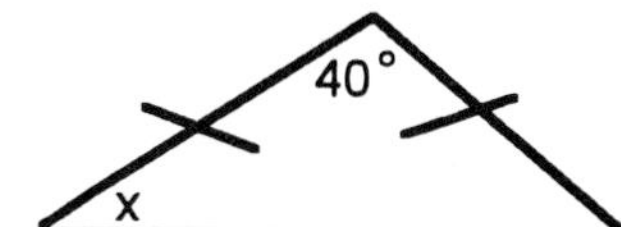

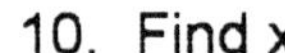

10. Find x

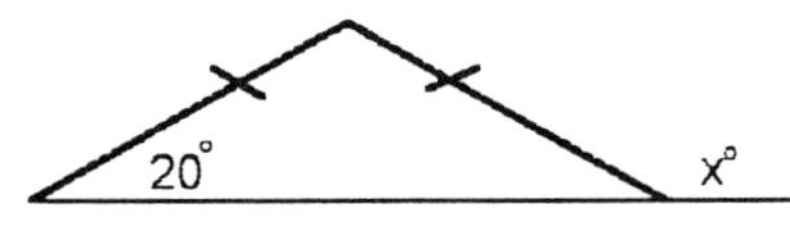

11. Find x

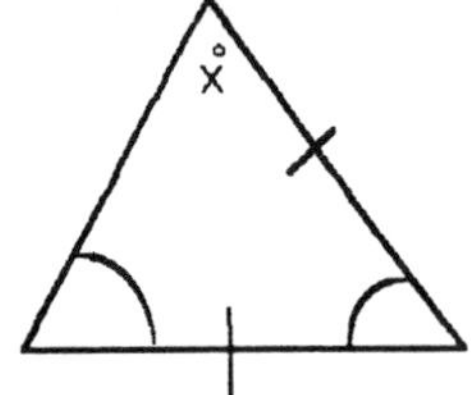

12. Find x

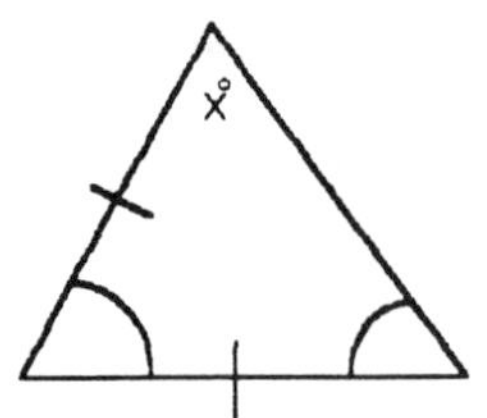

13. Find x

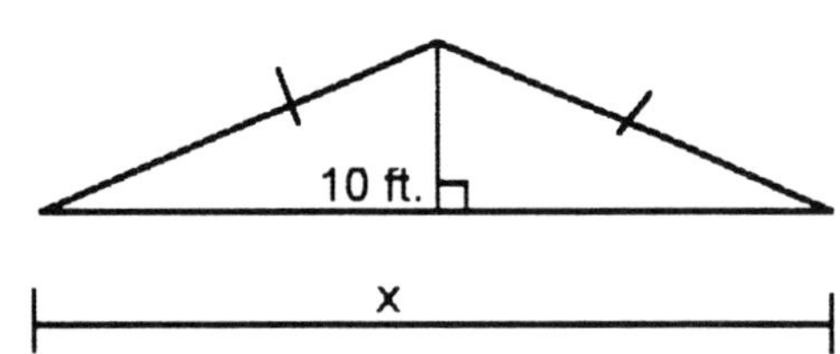

14. Find x

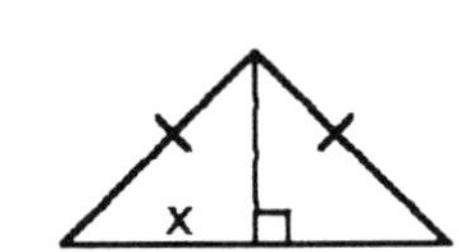

15. Find x

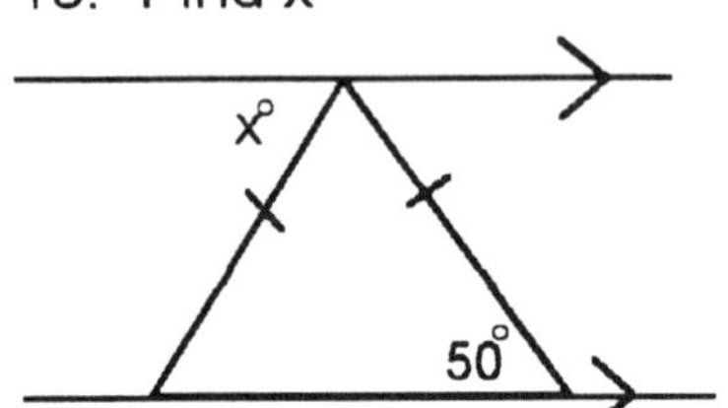

16. Find x

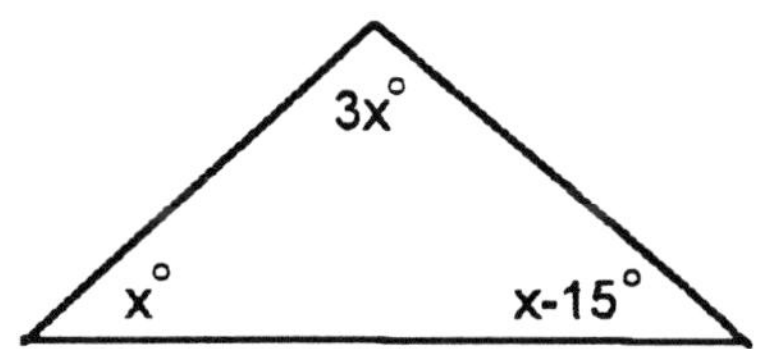

17. Find x

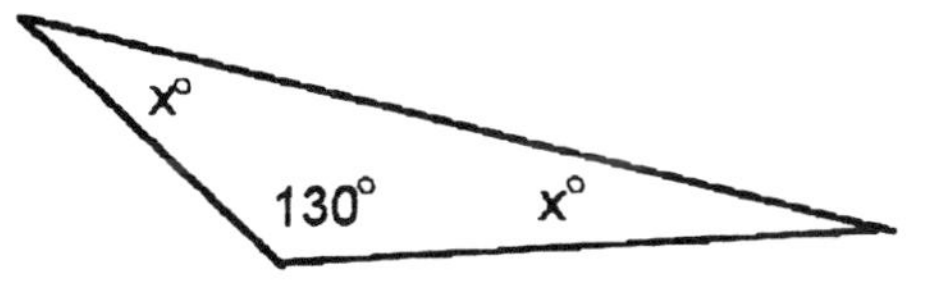

18. How many degrees are in each of the angles of an equilateral triangle?

19. Why are the acute angles of a right triangle complementary?

20. Why can there be at most one right or obtuse angle in any triangle?

21. What is the name of a polygon with three sides?

22. What is the name of a polygon with four sides?

23. If a triangle has three acute angles, what is it called?

24. What is the total number of degrees in a triangle?

25. A stop sign is what type of polygon?

26. What type of a triangle has two equal sides and two equal angles?

27. If a triangle has no equal sides, what is it called?

28. What is the name of a triangle with all equal sides?

2.3 Pythagorean theorem

In any right triangle, there is a special relationship between the sides. Special cases of this relationship were known by the Egyptians and the Chinese by at least 1000 BC, but it was in Greece in about 500 BC. that the relationship was formalized by Pythagoras.

The Pythagorean Theorem states: the square of the length of the hypotenuse of a right triangle is equal to the sum of the squares of the lengths of the other two sides. It is traditional to use c for the length of the hypotenuse, and a and b for the other two legs (sides).

Figure 9 - the Pythagorean Theorem

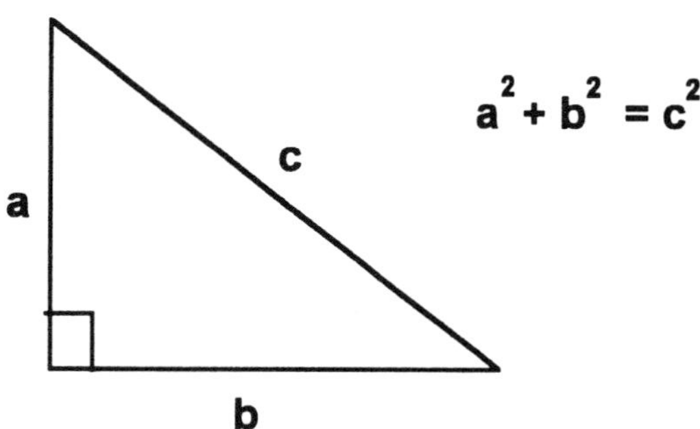

Here are some examples of the using the Pythagorean Theorem to find the missing side of a triangle:

Figure 10 - Pythagorean examples

a)

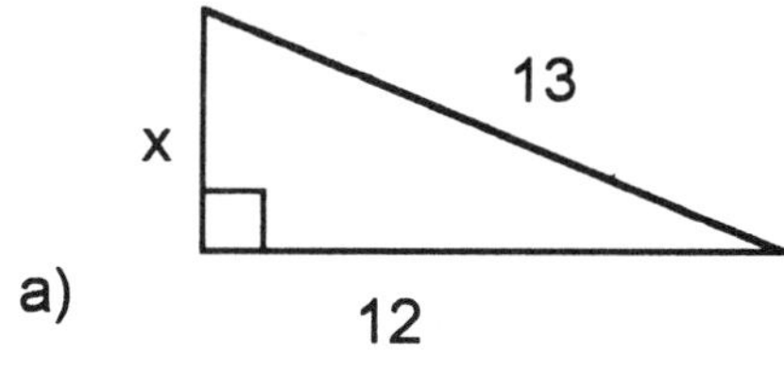

b)

8
x
15

In **(10a)**, length of the leg is missing.		In **(10b)** the hypotenuse is missing.
Theorem:	$a^2 + b^2 = c^2$	$a^2 + b^2 = c^2$
Substitute the values	$x^2 + 12^2 = 13^2$	$8^2 + 15^2 = x^2$
Square the values	$x^2 + 144 = 169$	$64 + 225 = x^2$
Add/Subtr. from sides	$x^2 + 144 - 144 = 169 - 144$	$289 = x^2$
The result is:	$x^2 = 25$	$x^2 = 289$
Take the square root both sides	$\sqrt{x^2} = \sqrt{25}$	$\sqrt{x^2} = \sqrt{289}$
The solution to this equation is:	$x = \sqrt{25}$	$x = \sqrt{289}$
Simplifying the square root	$x = 5$	$x = 17$

Answer could be in simple radical form or as a decimal or *between* :

$x = \sqrt{12}$, $x = 2\sqrt{3}$ or $x = 3.464$ or $3 < x < 4$

Depending on what sort of answer is required of you, leaving the answer as a radical (the $\sqrt{\ }$ is the radical) may be appropriate, or you may need a decimal answer. Both 25 and 289 are perfect squares, and therefore both examples above have integer answers which can be obtained from a table of squares, table of square roots, or by calculator. The solutions to these problems would be $x = 5$ and $x = 17$.

If your answer is a number such as $\sqrt{162}$, and your table of square roots only goes up to 100, you need to simplify the radical. Prime factor the radicand (the 162) into $2 \cdot 3 \cdot 3 \cdot 3 \cdot 3$, and then write $\sqrt{2 \cdot 3 \cdot 3 \cdot 3 \cdot 3}$. For every pair of like factors under the radical, you bring out one number representing the pair, and then that number multiplies the radical. In this case, there are two pairs of 3's, so each of those pairs of factors inside the radical produces a single factor in front of the radical. The $\sqrt{2}$ can be found in the table of square roots (it is 1.41421 . . . , which should be rounded to whatever place is requested).

$\sqrt{162}$

$\sqrt{2 \cdot 3 \cdot 3 \cdot 3 \cdot 3}$

$3 \cdot \sqrt{2 \cdot 3 \cdot 3}$

$3 \cdot 3 \cdot \sqrt{2}.$

$9\sqrt{2}$

9(1.414)

12.726

Another way to handle this kind of problem is to have a list of perfect squares and find where 162 fits in the n^2 column, and then you know that $\sqrt{162}$ has to be in the same relative place in the n column. **See Table B for Table of Squares Roots.**

n	n^2	$\sqrt{n}$	n	n^2	$\sqrt{n}$	n	n^2	$\sqrt{n}$
1	1	1.000	8	64	2.828	14	196	3.742
2	4	1.414	9	81	3.000	15	225	3.873
3	9	1.732	10	100	3.162	16	256	4.000
4	16	2.000	11	121	3.317	17	289	4.123
5	25	2.236	12	144	3.464	18	324	4.243
6	36	2.449	**12.726**	**...162**		19	361	4.359
7	49	2.646	13	169	3.606	20	400	4.472

This tells us that $\sqrt{162}$ is between 12 and thirteen (and closer to 13 than to 12). In some cases, this may be all the accuracy you need.

Wrap-up: $\sqrt{162}$ as an answer may be, depending on what is requested of you: $\sqrt{162}$, $9\sqrt{2}$, 12.726, or "a number between 12 and 13".

You may have noticed that the two examples given above involved right triangles where all three sides were integers. When this happens, the three numbers are called a **Pythagorean Triple.** The smallest Pythagorean triples are:

3, 4, 5   5, 12, 13   8, 15, 17   7, 24, 25

Multiples of Pythagorean triples are themselves Pythagorean triples

6, 8, 10,   10, 24, 26   9, 12, 15   12, 16, 20

These triples tend to show up frequently in standardized tests - **learn them.**

⌘ **4** Find x for these triangles?

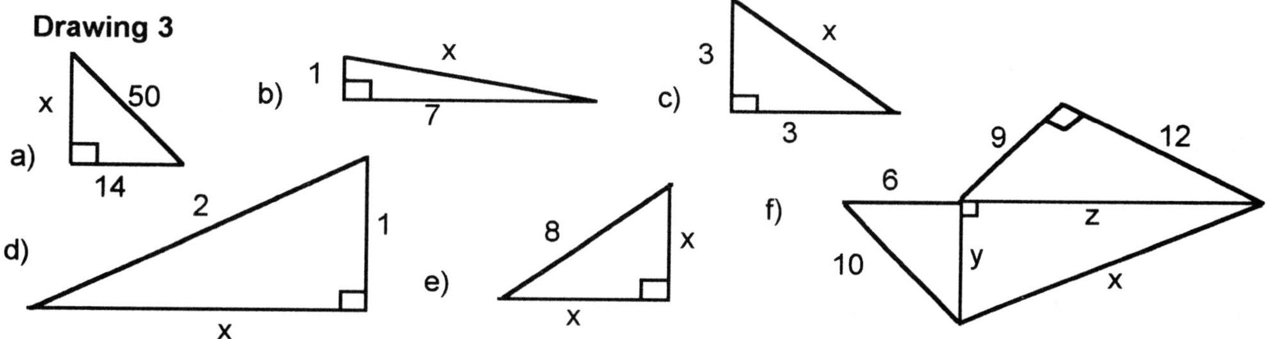

2.4 Quadrilaterals

The sum of the angles of a quadrilateral is 360. There are many possible quadrilaterals, but only special quadrilaterals usually appear in problems and applications. Most concerns about the general quadrilateral tend to be handled by dividing the quadrilateral up into triangles and applying trigonometry. The special quadrilaterals can be divided into two groups: parallelograms and trapezoids. The parallelogram group includes the parallelogram, the rectangle, the rhombus, and the square.

The **parallelogram** has two sets of parallel sides. It turns out that the diagonal connecting either pair of opposite vertices divides the parallelogram into two congruent triangles. As a consequence of this, the opposite sides of the parallelogram are equal and the opposite angles of the parallelogram are equal. Because of the parallel sides, consecutive angles around the parallelogram are supplementary **(11a).** A special kind of parallelogram is the **rectangle,** so called because all four angles are right angles (and therefore equal to each other).

All of the properties of parallelogram apply to the rectangle. The rectangle has equal diagonals **(11b).** The **rhombus** is another special parallelogram. It is sometimes called the diamond shape. The rhombus is a parallelogram with equal sides and perpendicular diagonals **(11c).** The last of the special parallelograms is the **square,** which has all four sides equal and all four angles equal. The diagonals of a square are equal and perpendicular to each other **(11d).**

Figure 11 - parallelograms

a) parallelogram

b) rectangle

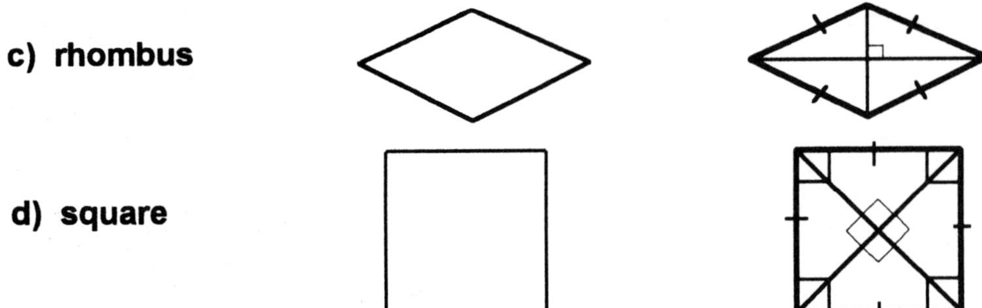

The **trapezoid** is a quadrilateral with one set of two parallel sides. **(12a,b)** Trapezoids may have two or three equal sides, or two right angles in them. **(12 d,e)** The only one of these with a special name is the **isosceles trapezoid. (12c)** The **median** of a trapezoid is the average of the lengths of the two parallel sides (called bases).

Figure 12 - trapezoids

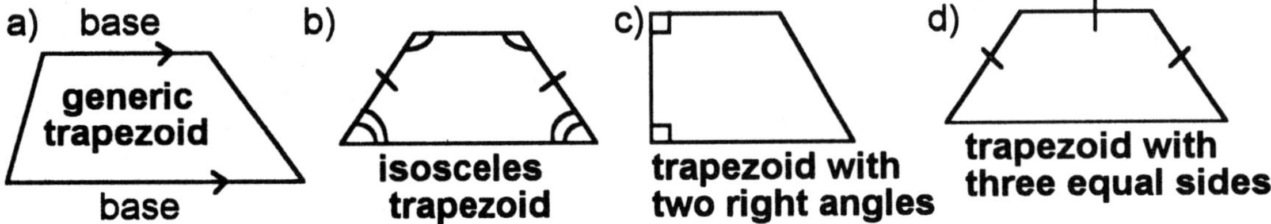

Exercises 2.3

Find the missing side. If it is a radical, express it
a) in simplified radical form
b) to three decimal places
c) as a number between two integers

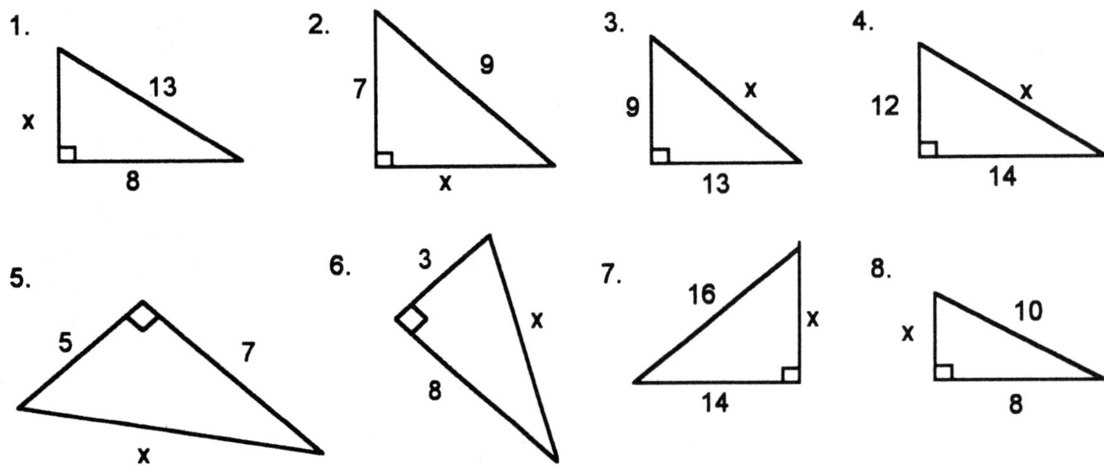

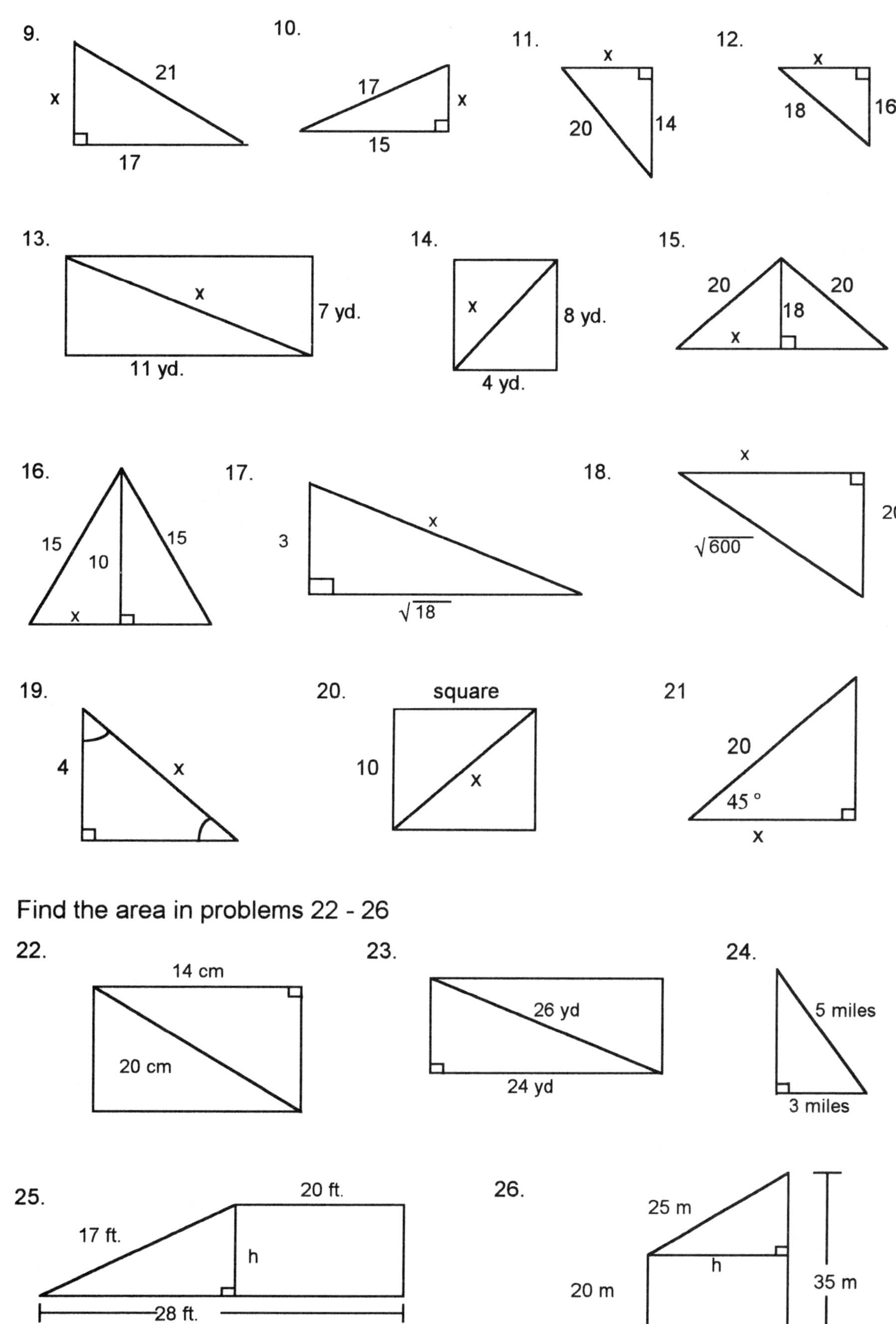
9.
x
21
17
10.
17
x
15
11.
x
20
14
12.
x
18
16
13.
x
7 yd.
11 yd.
14.
x
8 yd.
4 yd.
15.
20
20
18
x
16.
15
15
10
x
17.
3
x
√18
18.
x
20
√600
19.
4
x
20.
square
10
x
21
20
45 °
x
Find the area in problems 22 - 26
22.
14 cm
20 cm
23.
26 yd
24 yd
24.
5 miles
3 miles
25.
20 ft.
17 ft.
h
28 ft.
26.
25 m
h
20 m
35 m

Answers and some reasons to ⌘ problems: Chapter 2

⌘ 1 Answers will vary.

⌘ 2 (2a) The acute angles of the right triangle are complementary, so
$x + 67° = 90°$
$x = 23°$

(2b) The acute angles of an isosceles right triangle are equal and complementary, so $y = z$ and
$y + z = 90°$; **$y = 45°$; $z = 45°$**
x is the supplement of z
so $x + z = 180°$
$x + 45° = 180°$
$x = 135°$

⌘3 (3a) $40° + 100° + x = 180°$
$x = 40°$

(3b) $\angle 1 = x$ (isosceles triangle)
so $x + x + 20° = 180°$
$2x = 160°$.
$x = 80°$

(3c) (exterior angle principle)
$130° = 70° + x$
$x = 60°$
The other way to do this is:
(supplementary angles)
$130° + \angle 1 = 180°$
$\angle 1 = 50°$
Then $50° + 70° + x = 180°$
$x = 60°$

(3d) (isosceles triangle)
$\angle 1 = 40°$, and
$x + 40° + 40° = 180°$
$x = 100°$

(3e) The third angle principle guarantees that $x = y$.

⌘4 (4a) $x^2 + 14^2 = 50^2$
If you recognize that this is a multiple of the 7-24-25 right triangle, then the answer is easy: $2 \cdot 24 = 48$. Otherwise,
$x^2 + 196 = 2500$
$x^2 = 2304$
Answer: $x = \sqrt{2304}$
or: $x = 48$.

(4b) $1^2 + 7^2 = x^2$
$1 + 49 = x^2$
$50 = x^2$
Answer: $x = \sqrt{50}$,
or: $x = \sqrt{2 \cdot 5 \cdot 5} = 5\sqrt{2}$
or: $x = 7.070$

(4c) $3^2 + 3^2 = x^2$
$9 + 9 = x^2$
$18 = x^2$
Answer: $x = \sqrt{18}$,
or: $x = \sqrt{2 \cdot 3 \cdot 3} = 3\sqrt{2}$
or: $x = 4.242$

(4d) $1^2 + x^2 = 2^2$

$1 + x^2 = 4$

$x^2 = 3$

Answer: $x = \sqrt{3}$

or: $x = 1.736$

(4e) $x^2 + x^2 = 8^2$

$2x^2 = 64$

$x^2 = 32$

Answer: $x = \sqrt{32}$

or: $x = \sqrt{2 \cdot 2 \cdot 2 \cdot 2 \cdot 2} = 2 \cdot 2 \cdot \sqrt{2} = 4\sqrt{2}$,

or: $x = 5.656$

(4f) Note that the 10 - y - 6 triangle HAS to be a right triangle (section 1.4). In this problem you first have to find y and z, and then use y and z to find x.

i) $6^2 + y^2 = 10^2$

$36 + y^2 = 100$

$y^2 = 64$

$y = 8$

ii) $9^2 + 12^2 = z^2$

$81 + 144 = z^2$

$225 = z^2$

$z = 15$

iii) $8^2 + 15^2 = x^2$

$64 + 225 = x^2$

$289 = x^2$

Answer: $x = 17$

CHAPTER 3 RELATIONSHIPS BETWEEN TRIANGLES

Topics in this chapter:

- Introduction to Similarity and Congruence
- Proportions
- Similar Triangles
- Congruent Triangles

Answers to ⌘ try-out exercises are given at end of the chapter.
(3b) means that this item or idea is illustrated in **Figure 3, part b.**

3.1 Introduction to Similarity and Congruence

● **Two similar figures have the same shape, but are not generally the same size.** Being the same shape means that they must have equal corresponding angles. In addition to corresponding angles, similar figures have proportional sides - corresponding sides form a proportion. This is an extended proportion containing several ratios, with the side lengths from one figure forming the numerators and the side lengths from the other figure forming the denominators. (Ratios and proportions will be defined and applied in **3.2.**) Each of these ratios is equal to the same number, called the **scale factor,** k.

● **Two congruent figures are two figures which have all corresponding parts equal** - equal sides, equal angles, and equal interior segments, and in the same order. The figures may not be positioned in the same way, but they are in all other respects identical.

In short, congruence and similarity can be summarized as

**SIMILAR MEANS SAME SHAPE, DIFFERENT SIZES.
CONGRUENT MEANS IDENTICAL SIZE AND SHAPE.**

Quadrilateral ABCD is congruent to quadrilateral A'B'C'D' is written in symbols as **ABCD ≅ A'B'C'D'.** Three comments about this notation:

1) It is very common to imply corresponding angles by using the X - X' notation for corresponding vertices. In this case, this implies that **A** corresponds to **A'**, **B** to **B'**, etc. This also means that great care should be taken when naming a pair of congruent figures which have

different letters for vertex names to be <u>sure</u> that the vertices are named in corresponding order.

2) When there is a special little symbol for that polygon, such as the Δ for triangle, that symbol is used before the letters. When there is no such symbol (there is none for quadrilateral), just the vertices in proper corresponding order are used.

3) The mere statement that **ABCD $\cong$ A'B'C'D** immediately asserts that there are 4 equal corresponding sides and 4 equal corresponding angles **(1b).**

Quadrilateral ABCD is similar to quadrilateral A'B'C'D' is written in symbols as **ABCD $\sim$ A'B'C'D'**. Note that the similarity symbol **~** is the tilde. If you compare the similarity symbol (~) with the congruence ($\cong$), you will notice that the tilde is in both, but the similarity symbol has no = in it, because in similarity you are not asserting anything about equal size. The same types of notation facts listed above for congruence are true for similarity, except that in 3). Similarity implies only equal angles and nothing about equal size **(1a).** The following figures show what similarity and congruence mean for quadrilaterals.
Similar and congruent triangles will be considered in detail in sections 3.3 and 3.4.

Figure 1

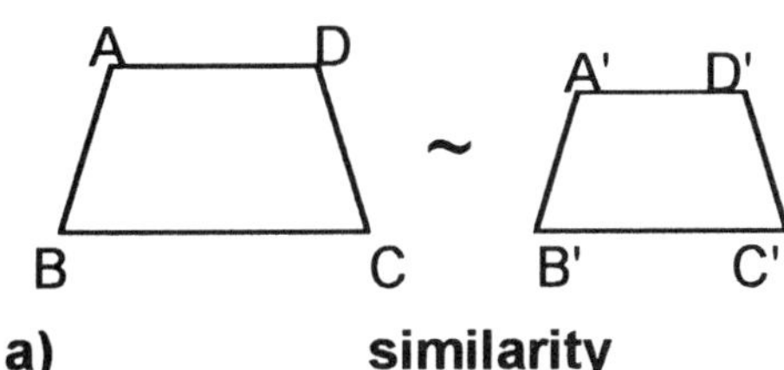

a) similarity

ABCD $\sim$ A'B'C'D'

equal angles

$\angle A = \angle A'$ $\quad$ $\angle C = \angle C'$

$\angle B = \angle B'$ $\quad$ $\angle D = \angle D'$

proportional segments

$$\frac{AD}{A'D'} = \frac{DC}{D'C'} = \frac{CB}{C'B'} = \frac{BA}{B'A'}$$

b) congruence

ABCD $\cong$ A'B'C'D'

equal angles

$\angle A = \angle A'$ $\quad$ $\angle C = \angle C'$

$\angle B = \angle B'$ $\quad$ $\angle D = \angle D'$

equal segments

$\overline{AD} = \overline{A'D'}$ $\quad$ $\overline{CB} = \overline{C'B'}$

$\overline{DC} = \overline{D'C'}$ $\quad$ $\overline{BA} = \overline{B'A'}$

⌘ 1 a) If ACDRF ~ ZXWTF, then $\angle d = \angle$**?**, $\frac{AR}{FD} = \frac{ZT}{?}$

b) If ΔAKZ $\cong$ ΔDWB, then ZK = **?**, $\angle W = \angle$**?**

3.2 Proportions

A proportion is an equality between two ratios. A ratio is a method of comparing two numbers. The ratio of a to b can be written as a : b, or in

fractional form as $\frac{a}{b}$. Although the ratio looks just like a fraction and manipulates just like a fraction, it means something different and is used for different things.

- **The fraction describes a part out of a whole.** The fraction $\frac{3}{8}$ means that you are talking about 3 out of the 8 equal parts of a whole. For example, if a pizza is cut into 8 equal pieces then we say the whole pizza has 8 equal parts. If you have eaten 3 of those 8 pieces, then you have eaten $\frac{3}{8}$ of the pizza.
- **A ratio compares one thing to another without reference to the whole.** The ratio 5 to 4, or $\frac{5}{4}$ means 5 of one sort of thing compared to 4 of another sort of thing. For example, you might examine the toppings on a single piece of pizza and note that it had 5 pepperoni and 4 black olives. You could then state that your piece of pizza had pepperoni to black olives in a 5 to 4 ratio.
- You could just as meaningfully turn the **ratio** upside-down, as long as the associated words are also turned around, and say that your black olive to pepperoni ratio was 4 to 5. The **fraction** $\frac{3}{8}$ cannot be meaningfully be turned upside-down - does it make any sense to say that you ate 8 out of the 3 pieces?
- There is a "whole" hidden in every ratio. Suppose you had a volleyball team composed of 4 women and 5 men. The ratio of men to women is 5:4. If you ever needed to get the ratio of women to the entire team (the "whole"), first you add men + women = 9 = entire team. Then the ratio of women to the entire team would be 4:9.

Ratios are frequently encountered in cooking, or in any process which requires mixing things. Your car's manual might specify that the proper antifreeze/water mixture should be 3 parts water to 2 parts antifreeze. This means that if you were to use 2 quarts of antifreeze, you would need to mix in 3 quarts of water, or 7.5 pints of water would require 5 pints of antifreeze. In each case the ratio could be reduced (just like a fraction) to $\frac{3}{2}$.

A proportion is an equality between two ratios. A proportion is used when you know a ratio relating to two quantities is true, and you are wish to apply the ratio to find out a missing quantity. Suppose the manual specifies using an antifreeze to water ratio of 2 to 3. You have 8 pints of antifreeze, and you want to mix in the right amount of water. How much water should you mix in? To solve this we need to realize a fact about proportions:

Every proportion has four sides: the top side, the bottom side, the left side, and the right side. As long as the proportion is set up so that each of the four sides match, the proportion is correctly set up to solve the problem.

Figure 2 shows to apply that rule to set up this problem.

Figure 2 - setting up a proportion with two equal ratios, same words.

$$\frac{\text{antifreeze}}{\text{water}} = \frac{\text{antifreeze}}{\text{water}}$$

	known things (given ratio)		unknown things (one of the amounts is unknown)		
antifreeze	2	=	8 pints	antifreeze	top side
water	3		x	water	bottom side
	left side		right side		

We need to say a few things about this setup:

- It is not unique. There are actually four ways to correctly set up this proportion, and four ways to set it up incorrectly. Another correct setup is $\frac{2}{8 \text{ pints}} = \frac{3}{x}$, because it might be easier to think

  $$\frac{\text{ratio part of antifreeze (2)}}{\text{pints of antifreeze (8)}} = \frac{\text{ratio part of water (3)}}{\text{unknown pints of water (x)}}.$$

  You need to pick a method of setting up the proportion and stick with it.

- **The order of the parts in ratios and proportions is critical.** For example, if you were to place a bet where the odds were 20 to 1, and you won, would you accept being paid on a 1 to 20 basis? If a recipe says 1 cup of sugar to 19 cups of flour, would you like to eat the result of using 19 cups of sugar to 1 cup of flour?

- Another way to approach the problem is to express the facts in two sentences, being quite careful to maintain a "corresponding order". For example, if the problem is :

  "If the sales tax on a $21.50 purchase is $1.50, what would the tax be on a $7.68 purchase?"

  You need to be careful here, because all four amounts will be in dollars.

  You would get no clue from the difference in units, as we would in a water/antifreeze problem.

  Rewrite the information in two sentences such as:

  $1.50 sales tax on a $21.50 purchase
  how much sales tax on a $7.68 purchase

  which becomes

  $$\frac{\$1.50 \text{ sales tax}}{\$ \text{ how much tax}} \; \frac{\text{on a}}{\text{on a}} \; \frac{\$21.50 \text{ purchase}}{\$7.68 \text{ purchase}} \rightarrow \frac{\$1.50}{x} = \frac{\$21.50}{\$7.68}$$

- The ratio 2:3 is a **unitless ratio.** That means that it has no units such as feet or hours in it. A unitless ratio can be applied to any same-units type of situation. For example, if you have to use a 3 to 4 ratio of water to flour, it does not matter if you use 3 pints of water and 4 pints of flour, 3 cups of water and 4 cups of flour, or 3 liters of water and 4 liters of flour.

- If you ever start out with a ratio which must be simplified before you can use it, it will be one of three possible situations.

  a) **unitless ratio** - simplify it. Example: 3 to 0.5 or $\frac{3}{0.5}$. This can be simplified by $\frac{3}{0.5} \cdot \frac{10}{10} = \frac{30}{5} = \frac{6}{1}$, or by dividing 0.5 into 3 to get 6 The answer should be left as $\frac{6}{1}$ because the ratio is the only fraction where you should leave the denominator of **1** in the final answer.

  b) **the ratio has units in the numerator and in the denominator which are the same sort of category** (time, distance, weight, area, capacity, money, jobs, etc.). Transform one of the units into the other, or both into a third kind of units. Then reduce, including canceling the units. The result should be a unitless ratio. Example: 8 dimes to 2 quarters - both are money. 8 dimes = 80 cents, 2 quarters = 50 cents, and the ratio becomes $\frac{80 \text{ cents}}{50 \text{ cents}} = \frac{8}{5}$, a unitless ratio.

  c) **the ratio has units which are not the same sort of category** (miles per gallon, kilometers per meter, gallons per square foot, dollars per square yard, etc.) This kind of ratio is called a **rate.** The 2 parts antifreeze to 3 parts water problem was this sort. Reduce the fraction and keep the units. **To find a unit rate, divide the numerator by the denominator for the unit rate's numerator, and put a 1 in the denominator.** Then the units may

  be written "numerator-unit per denominator-unit." Example: To change 60 miles in, to, or per, 3 hours into a unit rate, take $\frac{60 \text{ miles}}{3 \text{ hours}}$ and divide the 3 into the 60 to get $\frac{20 \text{ miles}}{1 \text{ hour}}$, which is 20 miles per hour, or 20 mph.

In any case, you should always end up with a positive whole number in the denominator.

Solving a proportion, once it is correctly set up, is often done by a method called cross-multiplication.

If you have a proportion, $\frac{a}{b} = \frac{c}{d}$,

the next step is **ad = bc.** Then solve for your unknown.

For example, to solve the proportion set up in the sales tax problem above,

Set up proportion: $\frac{1.50}{x} = \frac{21.50}{7.68}$

Then cross-multiply: $(21.50) \cdot x = (1.50)(7.68)$.

Then $21.50x = 11.52$

And then $x = 0.5358$

Or (rounding) **$x = \$0.54$.**

⌘ **2** a) Simplify: 16 dimes to 5 quarters b) Simplify: 144 to 24

c) Simplify: 130 square feet per 26 gallons

d) Write as a proportion equation: 3 is to 5 as 9 is to 15

e) Solve: $\frac{3}{4} = \frac{x}{24}$

f) Solve: If 25 shirts cost $640, how many shirts could you buy for $192?

g) Solve: An adult ed class of 40 students has 22 women in it. At that same rate, how many men would be in a different class with 140 in it?

EXERCISES 3.1 & 3.2

1. If a railroad enthusiast builds a model of a steam engine where $\frac{1}{4}$ inch in the model represents $1\frac{1}{4}$ feet in reality. Express the scale factor as a simplified ratio.
2. A foreign exchange bank exchanges 250 yen for 20 dollars. Express this ratio in simplest form.
3. 420 gallons of water can be treated by 2.1 gallons of water purifier.. Express this as a unit rate.
4. A heavy truck can travel 520 miles on 39 gallons of gas. Express this as a unit rate.
5. Simplify the ratio of 2 quarters to 3 dimes.
6. Simplify the ratio of 9 days to 3 weeks.
7. The giant size of assorted candy has 500 pieces in it, including 125 pieces of taffy and 100 bonbons. The remaining candies are all hard candies. What is the ratio of taffy pieces to the entire assortment? What is the ratio of bonbons to hard candies?

8. The super-deluxe box of chocolates has 500 pieces: 50 pure chocolate, 125 soft-filled, 75 crunchy-filled, 75 caramel-filled, 50 nougat-filled, and the rest were chocolate-covered nuts. What is the ratio of caramel filled to nougat-filled? What is the ratio of pure chocolate to the entire collection? What is the ratio of chocolate-covered nuts to soft-filled candies?
9. If 2 pairs of shoes sell for eighteen dollars, how much would 5 pairs of shoes cost?
10. If Marta earns $360 in 8 days, how much could she earn in 6 days?
11. A 12-lb. boneless rib roast contains 30 servings of meat. How many servings would be provided by a 15-lb. boneless rib roast?
12. If 3 pounds of hamburger costs $3.80, what is the cost of 5 pounds?
13. A store owner expects to make a profit of $2 on a make-up kit that sells for $15. How much profit would she expect to make on the kit that sells for $50?
14. Inez works for a well-known cosmetic company. If she makes $6 for every $100 worth of the product she sells, what will she make if she sells $6000 worth of the product?
15. A home-owner pays property taxes of $400 per year. If taxes are figured at a rate of $1.20 for every $100 of assessed value, what is the assessed value of her home?
16. The sales tax on an item costing $80 is $6.35. What is the sales tax on an item costing $120?
17. A paint store sold three hundred gallons of paint in five months. At that rate, how much paint can the manager expect to sell in 7 months?
18. If 4 cans of spray paint are needed to paint three lawn chairs, how many cans are needed to paint 10 chairs?
19. Electricity and gas for the Fresquez home averages $156 every two months. What is the total energy bill for a year?
20. A security guard has an 11-month contract that pays $30,000. How much would she be paid in $4\frac{1}{2}$ months?
21. Arthur and Carlos are driving to college at a steady speed. The first 200 miles of the trip took $4\frac{1}{2}$ hours. How long would it take them to drive the last 500 miles? (do not worry about rest-stops or lunch stops!)
22. If there are 10 pot-holes every 15 miles on the average. How many pot-holes would you expect in 200 miles?
23. A tray of donuts uses 6 cups of flour and two and one-half cups of sugar. How much flour and sugar must the baker order to make forty trays of donuts?
24. The distance between two cities on a road map is 6 inches. If 1 inch represents 40 miles, what is the distance in miles between the two cities?
25. An architect drew plans for a public building using a scale of $\frac{1}{4}$ inch to represent 25 feet. How many feet would 3 inches represent?

26. If two cities on a map are 5 inches apart and the actual distance between them is 253 miles, what is the actual distance in miles between two cities $4\frac{1}{2}$ inches apart on the map?

27. A cook needs 5 gallons of punch for every 20 people. He starts out with 50 gallons of punch, and then finds that he will have 235 people. How much more punch does he need?

28. A carpenter will use an average of 5 pounds of nails on a 2000 square foot roof. He has 3 pounds on hand. How much more will he need to get to do a 2500 square foot roof? Nails are sold in pounds to the nearest tenth pound.

29. A park ranger doing a wildlife survey catches and tags 100 wild geese, and releases them back into nature. A month later she catches 300 wild geese, and finds that 18 of them are tagged. How big is the wild goose population?

30. An EPA official is concerned about the white bass population in Large Lake. She sets up a net and catches 150 white bass which were at least 6 inches long. She tags them and releases them back into the lake. The next day she catches 500 white bass over 6 inches long, and finds that 75 of them were tagged. What is the white bass population in the entire lake?

3.3 Similar triangles

Similar triangles are triangles with the same shape, but different sizes. The same shape means that the two triangles have corresponding angles equal. Although similar triangles have three sets of equal angles, all that it takes to guarantee similarity is equality of **two** pairs of corresponding angles. This is oftentimes referred to as the **Angle-Angle principle or AA principle for similar triangles.** (The third angle principle takes care of the equality of the third pair.) **(3)**

Figure 3 - similar triangles

Δ ABC ~ Δ A'B'C'

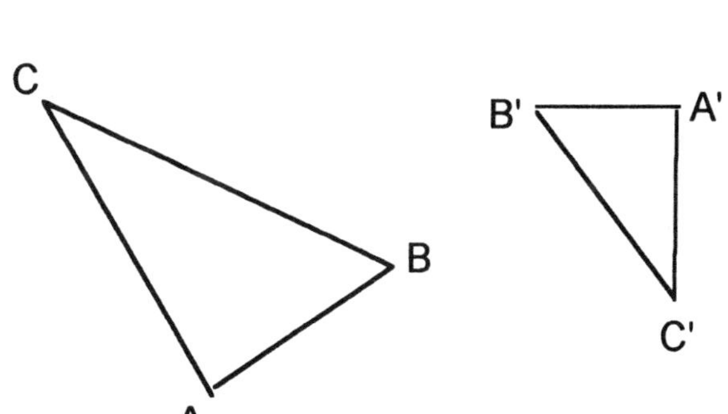

corresponding angles are equal

$\angle A \cong \angle A'$
$\angle B \cong \angle B'$
$\angle C \cong \angle C'$

and corresponding sides are in proportion.

$$\frac{AB}{A'B'} = \frac{AC}{A'C'} = \frac{BC}{B'C'}$$

Here are three examples of the application of similar triangles to find some missing quantities. Find the unknown quantities:

Figure 4 - applications of similar triangles

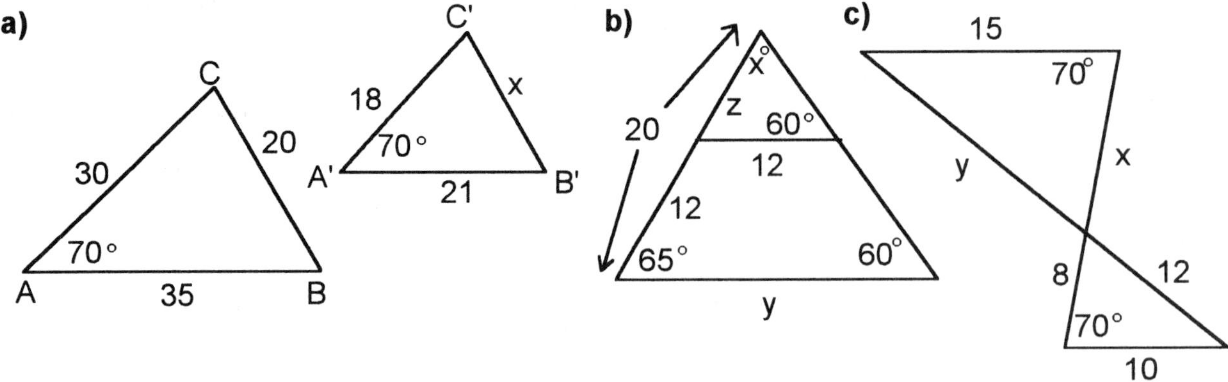

(4a) is straightforward. **(4b)** can be intimidating at first, and **(4c)** can be confusing, so it is important to find useful methods of attack

In **(4a)** ΔABC ~ ΔA'B'C' with AC corresponding to A'C', etc.

In **(4b)** there is actually one triangle inside another triangle, sharing the vertex at the top. The two triangles are similar, with the x angle at the top of both of the triangles, and the 60° angles matching up. In **(4'b)** below, we separate the two similar triangles.

In **(4c),** the figure is made from two similar triangles, with one of the triangles rotated with respect to the other as far as the corresponding parts are concerned. To find the corresponding angles, we trace around the triangles in corresponding order: 70° angle → unmarked angle → vertical angle in each triangle **(4'c).**

Figure 4' (4 revisited)

b) separating two similar triangles

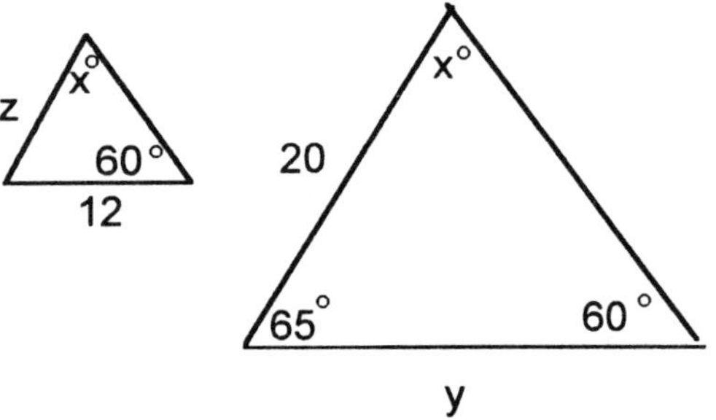

c) finding corresponding vertices

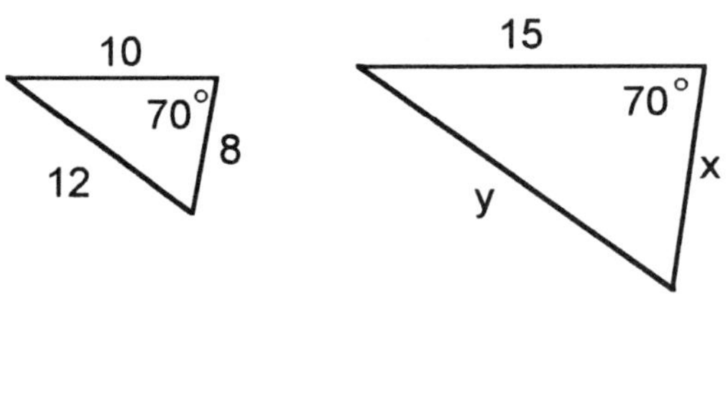

These techniques (separating overlapping triangles and tracing around the angles)can help to clear up confusion about what corresponds to what. Once you have the corresponding vertices determined, you can set up the appropriate proportions to find the unknown sides.

In **(4'a)** we first set up a proportion. The upper triangle will be the numerators, and the lower triangle will be the denominators.

	left side	right side		
upper triangle / lower triangle	$\frac{30}{18} = \frac{20}{x}$		or	$\frac{35}{21} = \frac{20}{x}$
cross-multiplying	$30x = (18)(20)$			$35x = (21)(20)$
	$30x = 360$			$35x = 420$
dividing	**$x = 12$**			**$x = 12$**

In **(4'b)** Two angles (the ones marked x, and the 60° angles) match, so the AA similarity principle guarantees similarity. The z and the 30 unit sides are corresponding, and the 12 unit and the y sides are corresponding. In the larger triangle, $x + 65° + 60° = 180°$, so **$x = 55°$**. The proportion set up for this problem is:

$$\frac{\text{left side of left triangle}}{\text{left side of right triangle}} = \frac{\text{bottom side of left triangle}}{\text{bottom side of right triangle}} \quad \text{or} \quad \frac{z}{20} = \frac{12}{y}$$

There is one other fact we need to make use of here: $20 = 12 + z$ (the left side of the larger triangle is made up of the 2 parts, one 12 units long, and the other z long). From this **$z = 8$.**

Then the proportion becomes	$\frac{8}{20} = \frac{12}{y}$
Using cross-multiplication	$8y = 20 \cdot 12$
Simplify	$8y = 240$
Answer	**$y = 30$.**

In **(4'c)**, the main thing is to make sure to use corresponding sides. Tracing around the triangles gives the side 12 units long in the upper triangle corresponding to the y side in the lower triangle, the side 10 units long corresponding to the side 15 units long, and the side 8 units long corresponding to the x side. Using these correspondences, the proportion shapes up as

$$\frac{12}{y} = \frac{10}{15} = \frac{8}{x}$$

Solving the two equations we obtain from it,

	$\frac{12}{y} = \frac{10}{15}$	$\frac{10}{15} = \frac{8}{x}$
Cross-multiplying	$12 \cdot 15 = 10y$	$10x = 8 \cdot 15$
Simplifying	$180 = 10y$	$10x = 120$
Answer	**$18 = y$**	**$x = 12$**

⌘3 Find x and y in the figures in **Drawing 1.**

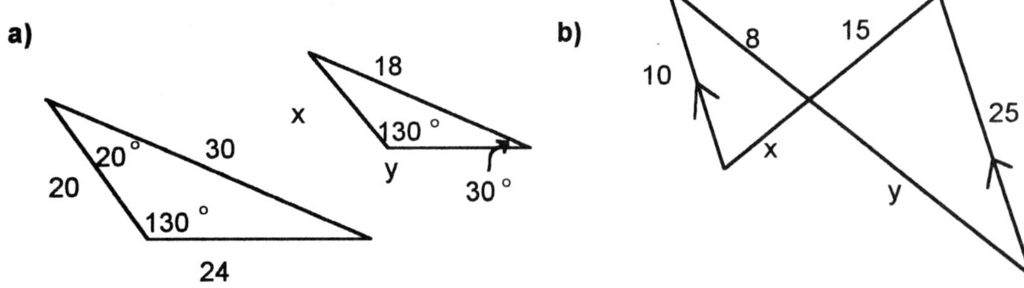

Let us see how similar triangles may be used in an application problem. Similar triangles are used frequently to obtain unknown heights and distances. The problem illustrated in **Figure 5** is: figure out how tall a pole is by using its shadow length, a man's height, and his shadow length. The assumptions made are that the man and the pole are both perpendicular to the ground, and that the sun at the same time makes the same angle to the ground for both shadows. Here is the procedure to set up proportions to solve the problem:

Figure 5 - using similar triangles to figure out height

At a certain time of day, a 6 ft. man casts an 8 ft. shadow, and a flag pole casts a 20 ft. shadow. How tall is the flag pole?

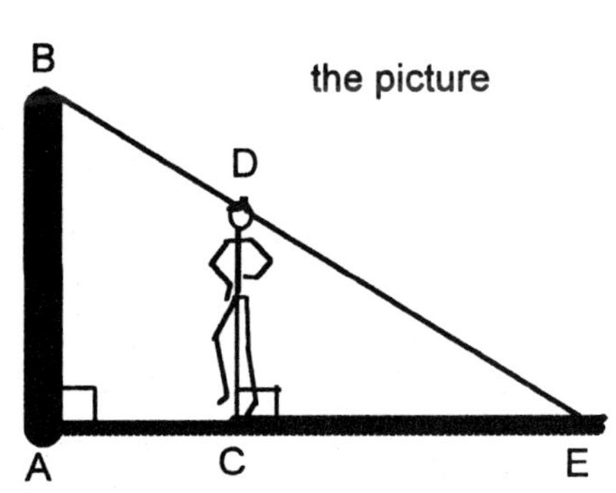

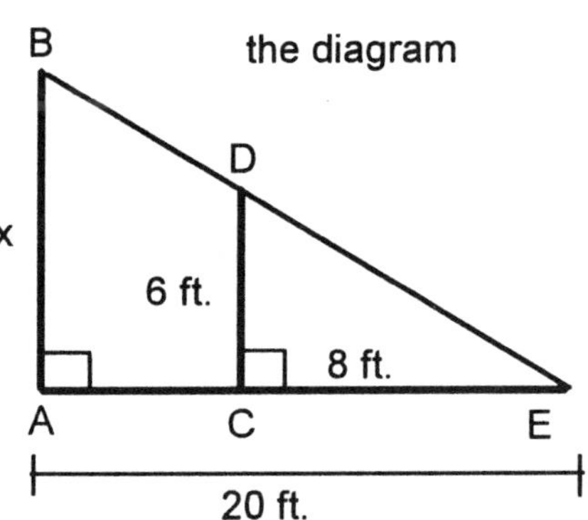

The fact that both shadows are made at the same time of day allows us to overlap the triangles. The sun will make the same angle with the ground in forming both shadows, therefore the right-most angle is the same so that $\angle$**E** is part of both triangles. We can say that the triangles are similar by the **AA** similarity principle ($\angle$**E** = $\angle$**E,** and the two right angles are equal).

Figure 6 - the triangle in (5) split into its two similar parts:

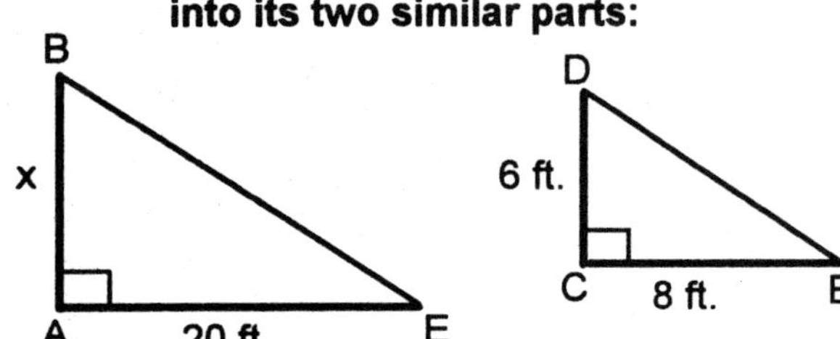

Setting up the proportion	$\frac{x}{6} = \frac{20}{8}$
Cross-multiplication	$8x = 6 \cdot 20$
Simplify	$8x = 120$
Answer	$x = 15$

The flagpole is 15 feet tall.

EXERCISES 3.3

The pairs of triangles are similar. Find the unknown quantities.

1.

9 "

21 "

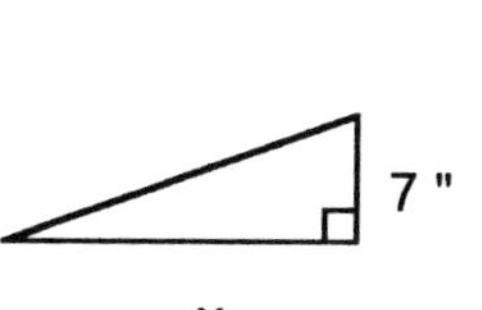

2.

15 m

x

12 m

9 m

3.

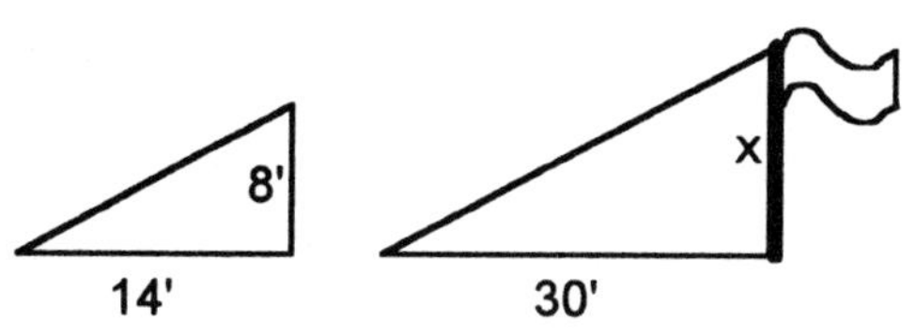

4.

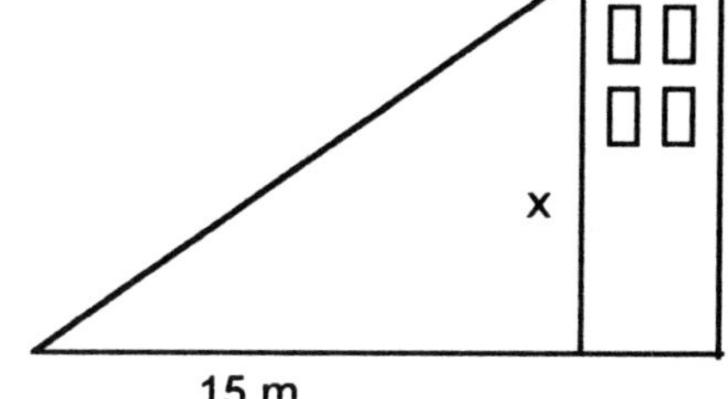

5.

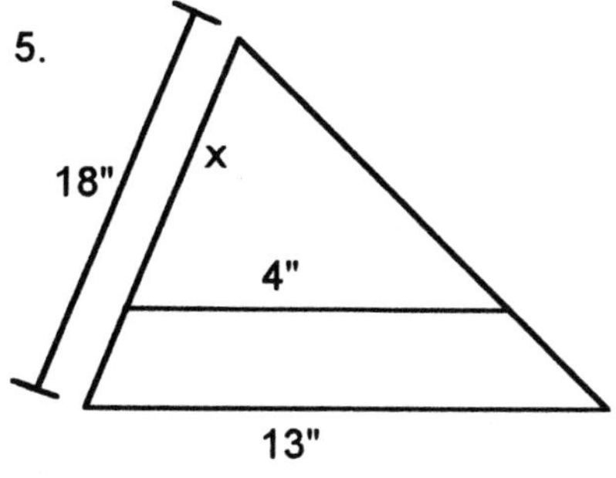

6.

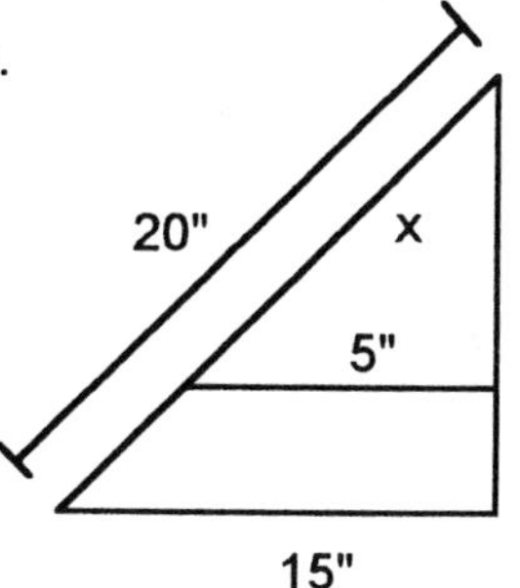

7.

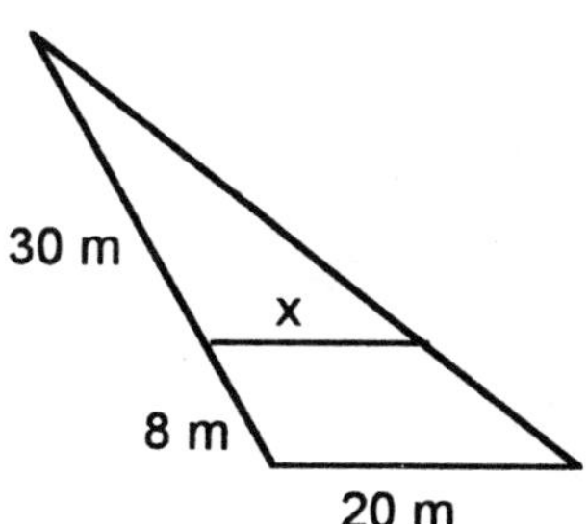

8.

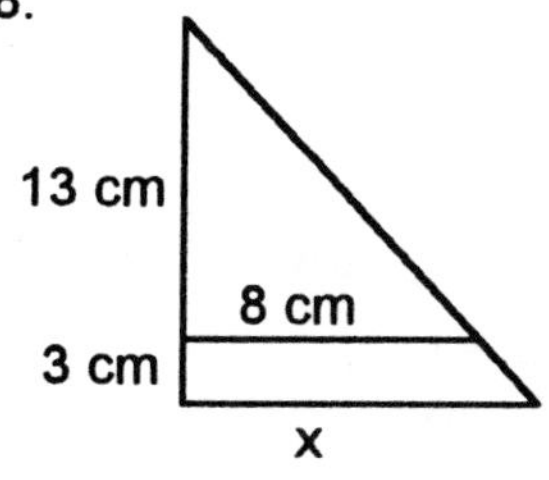

9.

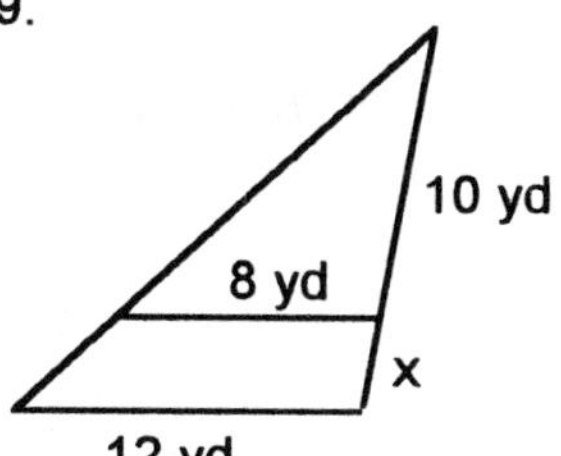

10.

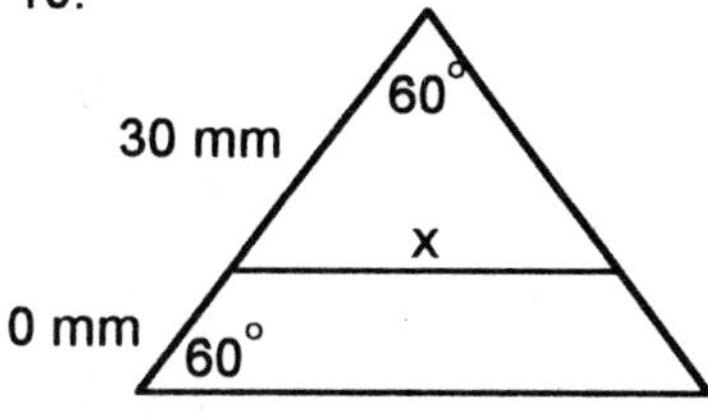

11.

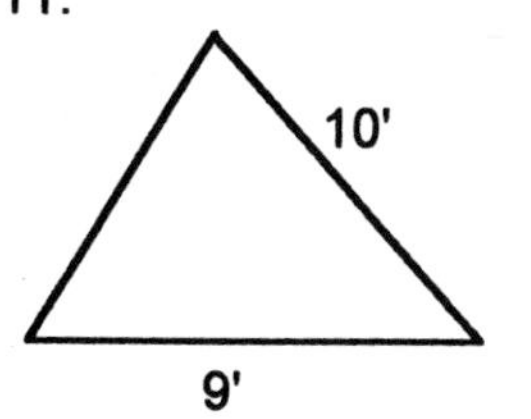

4'

x

12.

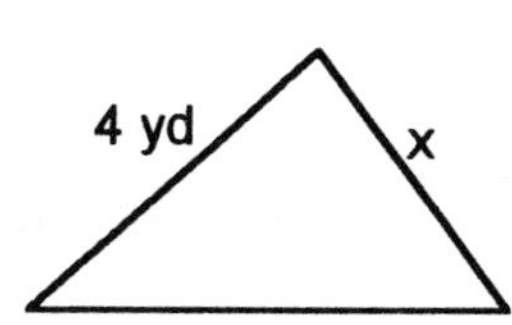

7 yd

5 yd

13. Find a and b

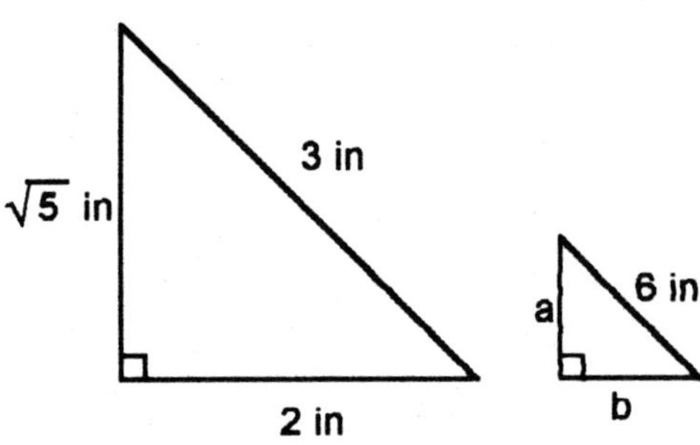

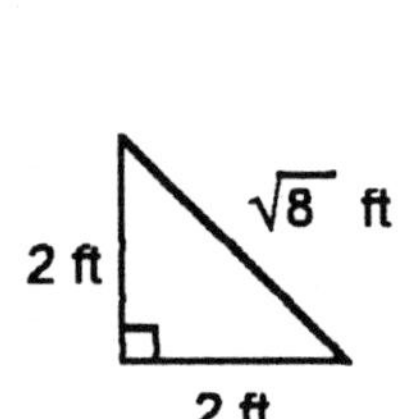

14. Find a and c

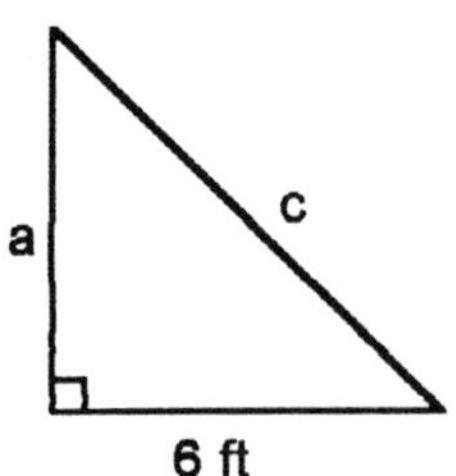

15.

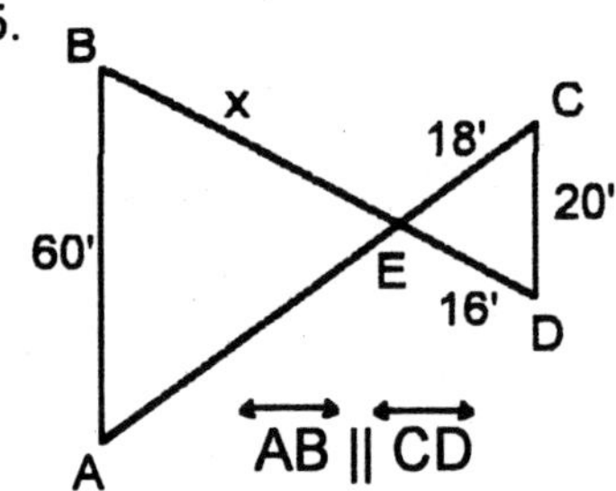

16.

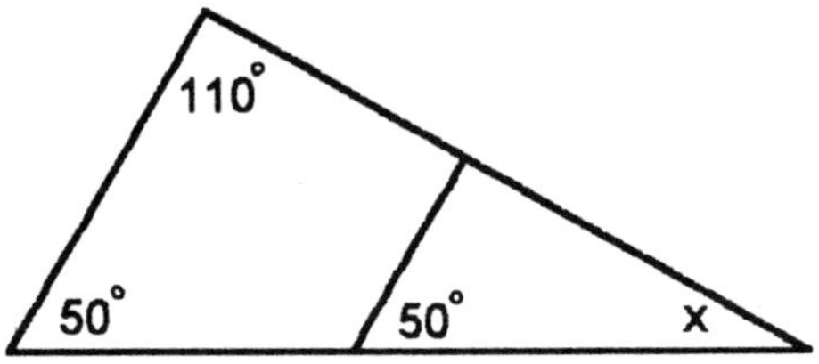

17. Find x

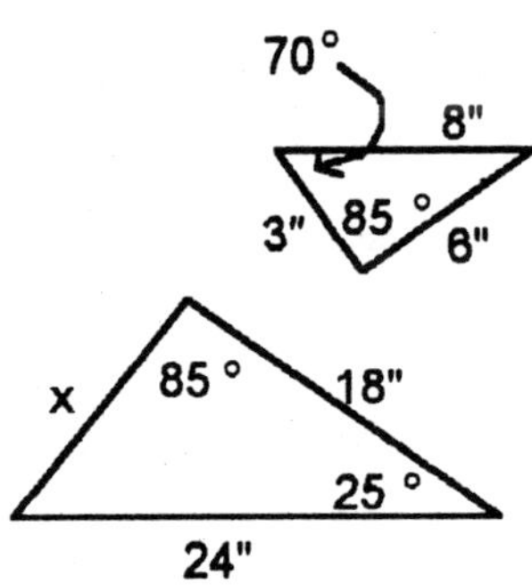

18. Find x, y, and z

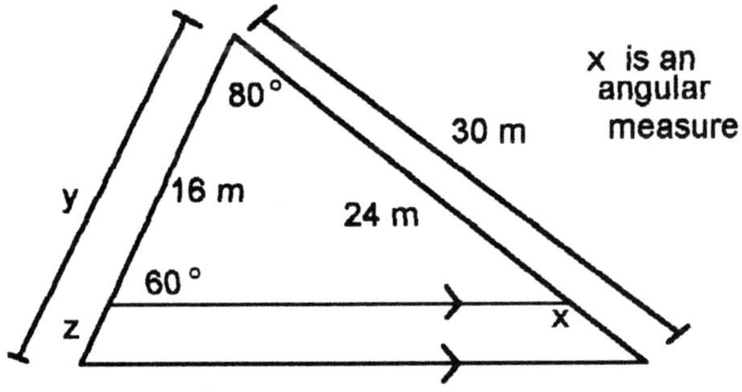

19. Find the distance across the lake (x)

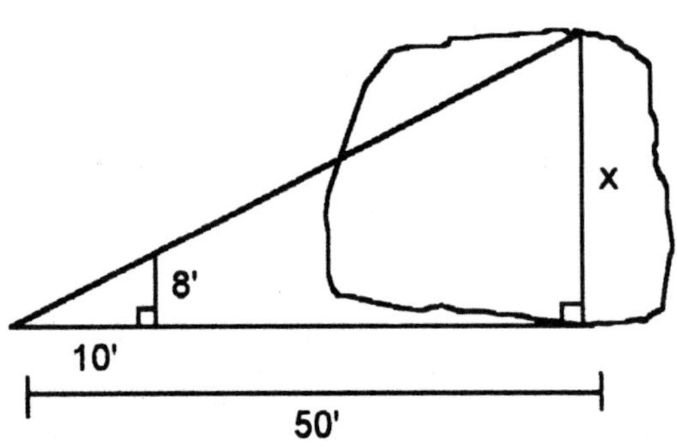

20. Find the distance across the island (x)

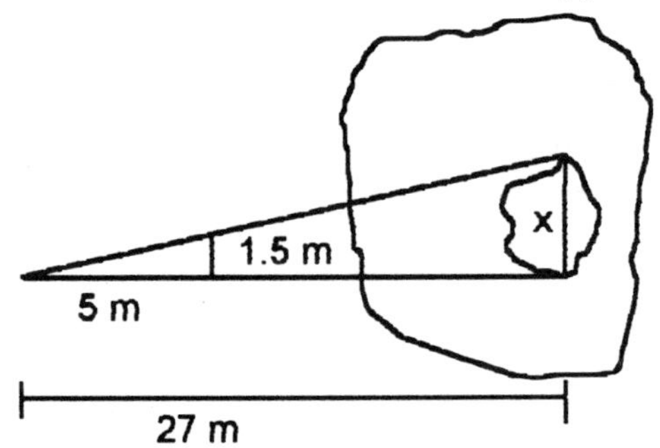

21. A 6-foot man has a 4-foot shadow when a tree has a 30-foot shadow. How tall is the tree?

3.4 Congruent triangles

There are two reasons for studying congruent (equal) triangles: sometimes you only need to be able to justify the statement that one triangle is congruent to another, but more often, you need to apply consequences of congruence to solve a problem. We will give a very basic presentation of the reasons for congruence and then hurry on to consider the applications.

If you are given two triangles congruent to each other, then, as a result, six things in one triangle are equal to the corresponding six things in the other triangle.**(8)** To go from given triangles to the fact that they are congruent only takes two or three corresponding things equal. To make it easy to remember, we have short names for these congruence principles: **SAS, SSS, ASA,** and **HL.**

Figure 7 - Congruent triangles

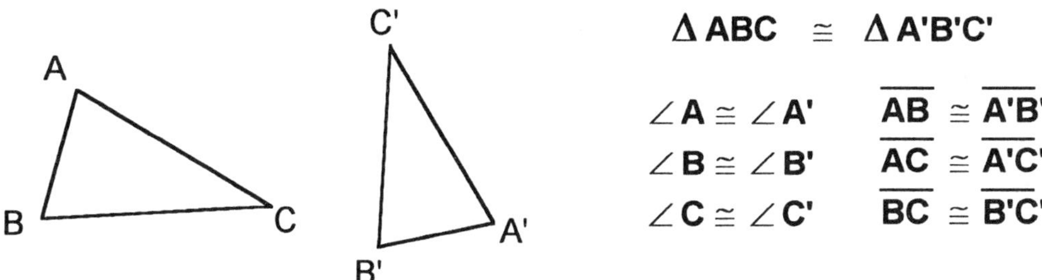

The **SSS principle of congruence** is that if all three sides of one triangle are equal to the sides of another triangle, the triangles are congruent (equal) **(8).**

Figure 8 - SSS principle of congruence

The **SAS principle of congruence** is that if two sides and the included angle of one triangle are equal to the two corresponding sides and the included angle of another triangle, then the triangles are congruent (equal) **(9a).** The "**included angle**" **(9b)** is critical, because if the equal angle is not included between the corresponding equal sides, there are in general two different triangles which may result **(9c).**

Figure 9 - SAS principle of congruence

a) $\cong$

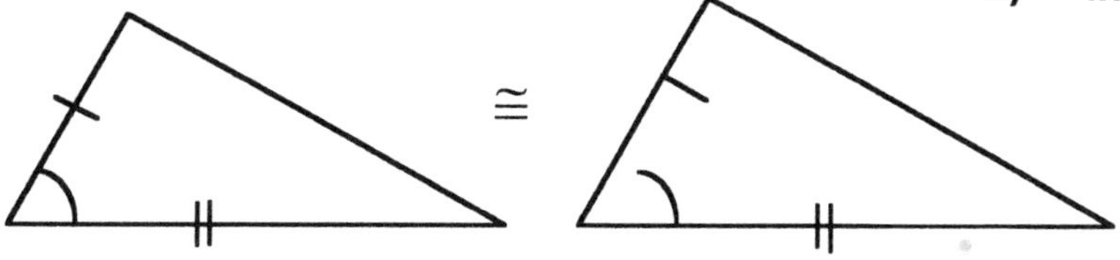

b) Included angle - - the sides are attached to the angle

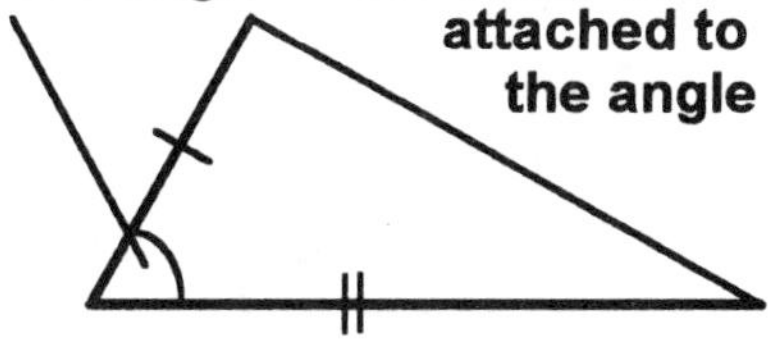

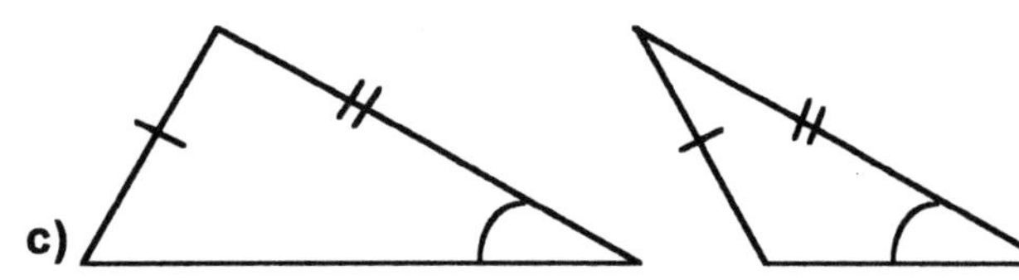

This is what can happen if the angle is NOT included - two triangles are possible, so you can not guarantee congruence.

The **ASA principle of congruence** is that if two angles and the included side of one triangle are equal to two angles and the included side of another triangle, then the triangles are congruent (equal) **(10a).** The "**included side**" **(10b)** is not crucial, because the Third Angle Principle will take the two-corresponding-angles-equal in the first triangle and two-corresponding-angles-equal in the second triangle and get the third corresponding angles equal. Then no matter which two angles you started out with, you actually have two corresponding equal angles which include the corresponding equal side in both triangles.**(10c).** This is sometimes named the **AAS** Principle. If you remember **ASA** and the Third Angle Principle you won't have to remember **AAS** separately.

Figure 10 - ASA principle of congruence

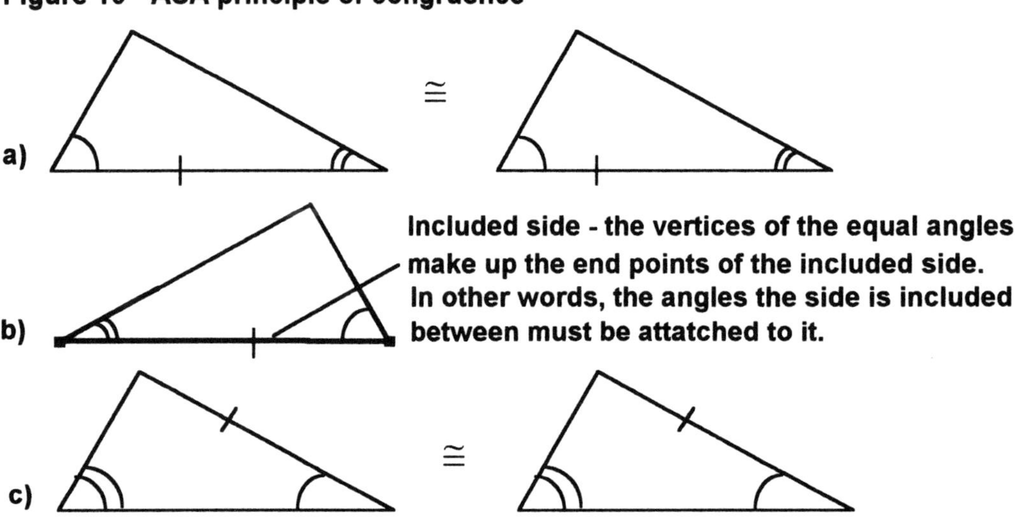

The third angle principle allows you to use ANY two angles with a side to get congruence. This is sometimes referred to as the AAS principle.

The **HL** principle of congruence is a special case - **it only applies to right triangles:** If two right triangles have equal hypotenuses and on pair of corresponding legs equal, the two triangles are congruent.

Figure 11 - HL congruence

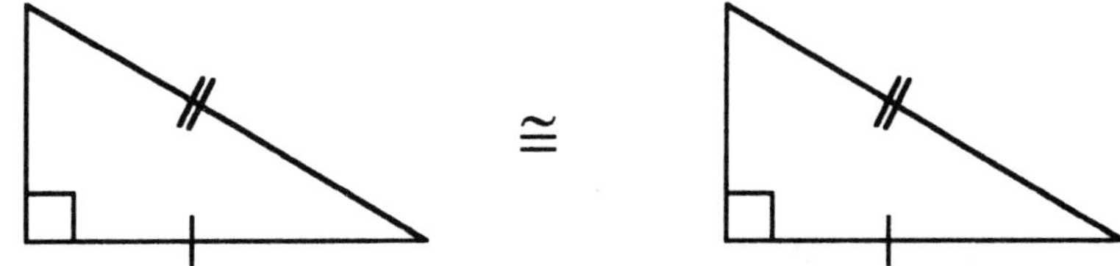

⌘ **4** For each pair of triangles, tell why they are congruent, and name the congruent triangles.

Drawing 2

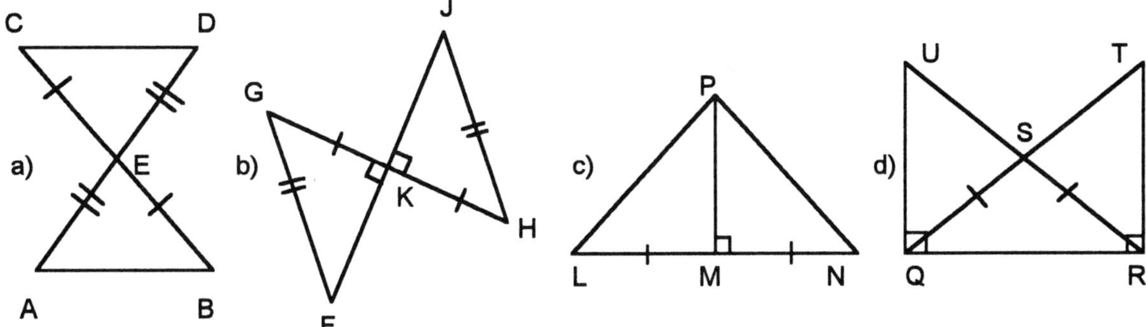

Applying congruence of triangles usually involves using the fact that once you have congruence, the corresponding parts of one triangle equal the corresponding parts of the other triangle. Recall the discussion about the isosceles triangle in section 2.2. The altitude to the base of the isosceles triangle divided the triangle into two congruent triangles. Let's see why:

Figure 12 - altitude to the base of an isosceles triangle

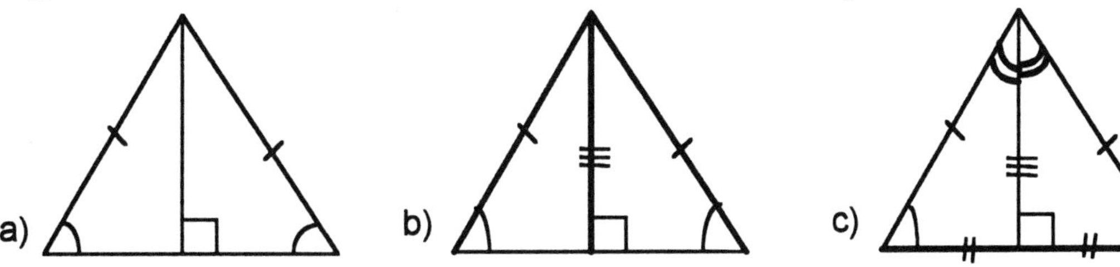

Isosceles triangle - comes with two equal sides and two equal angles already.	**The altitude is a common side, and therefore equals itself, so, by HL, the two triangles are congruent.**	**Congruence results : equal angles (parts of the vertex angle) and equal sides (parts of the base).**

⌘ **5** Find the values for x and y in Drawing **(3a)** and the value for x in Drawing **(3b).**

Drawing 3

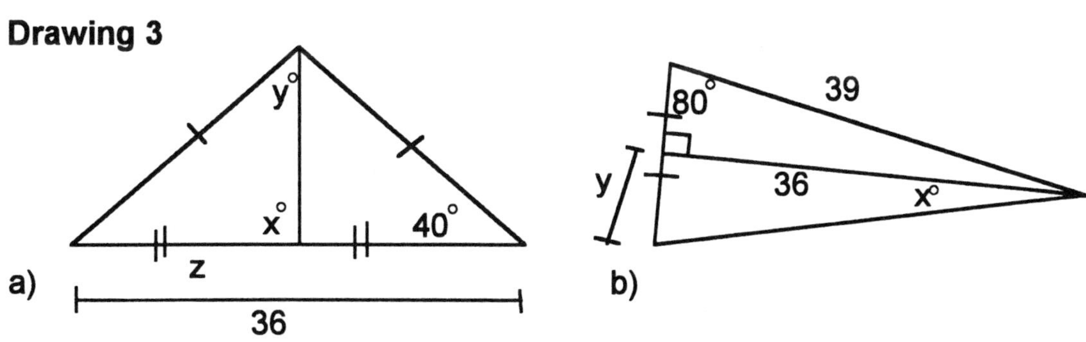

EXERCISES 3.4

1. $\overline{\mathbf{RQ}} \parallel \overline{\mathbf{MN}}$
   Show why
   $\Delta MNP \cong \Delta QRP$

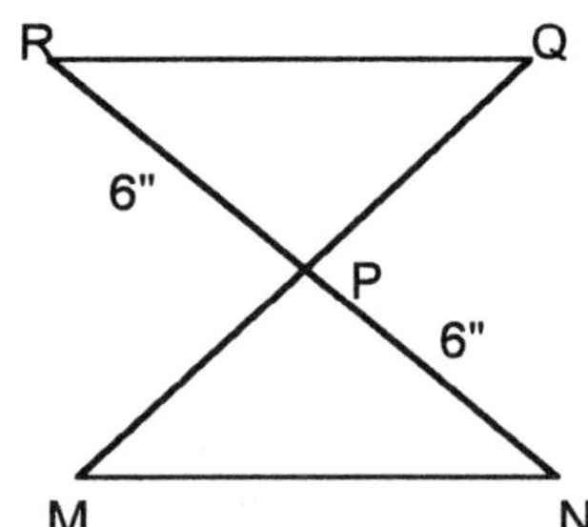

2. Show why
   $\Delta ABC \cong \Delta EDC$

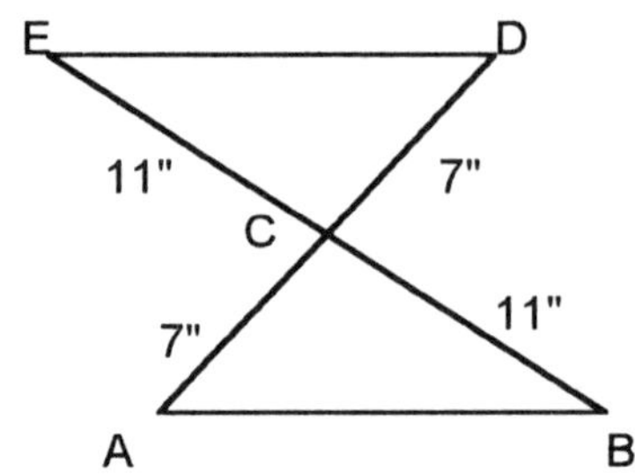

3. $\overline{\mathbf{KF}} \parallel \overline{\mathbf{JH}}$
   FK = JH, Show why
   $\Delta GFK \cong \Delta GJH$

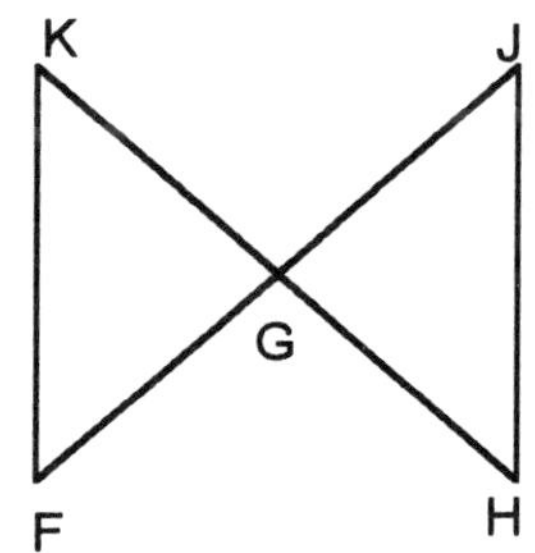

4 Show why
   $\Delta TPQ \cong \Delta RSQ$

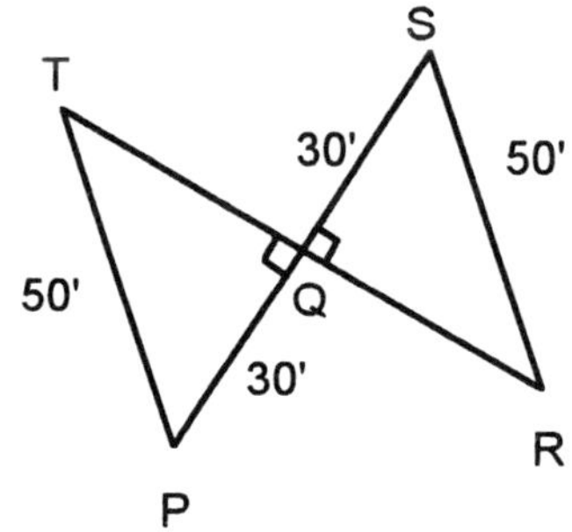

5. Show why
   $\Delta DBA \cong \Delta DBC$

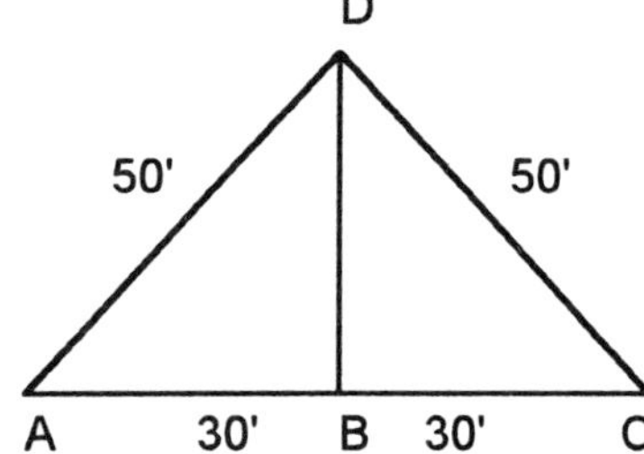

6. ΔPRS is isosceles
   $\overline{\mathbf{SQ}}$ is an altitude
   Show why $\Delta PQS \cong \Delta RQS$

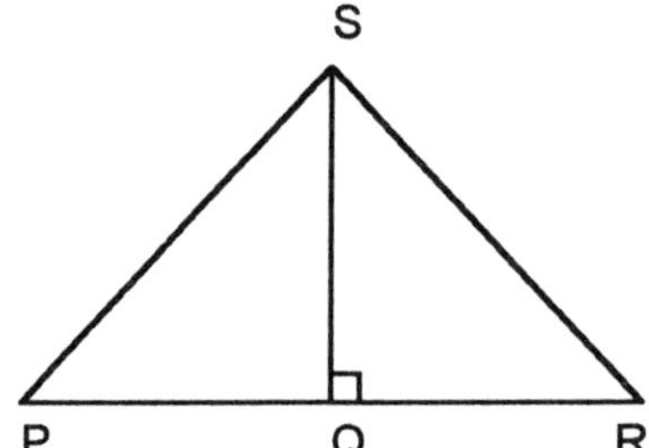

7. TQ = PQ = SQ = RQ
Show why
ΔSQR ≅ ΔTQP

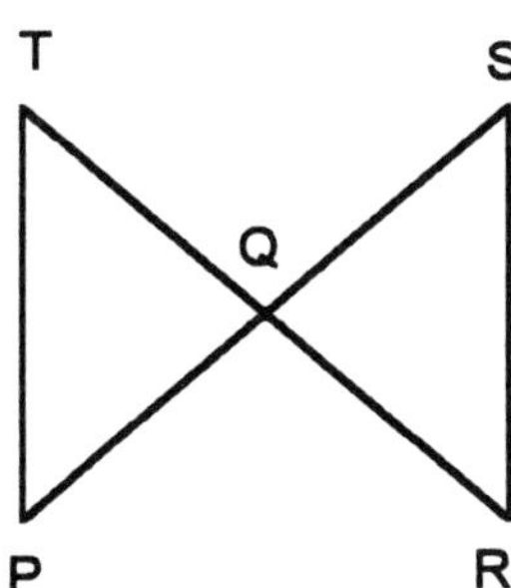

8. $\overline{FK} \parallel \overline{HJ}$, KG = GH
Show why
ΔGKF ≅ ΔGHJ

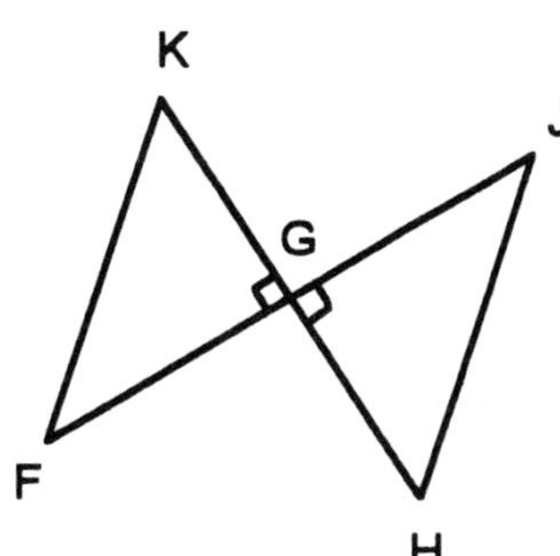

9. AD = DC
∠ADB = ∠BDC
Show why
ΔDBA ≅ ΔDBC

D

A B C

10. WX = XY, WZ = YZ
Show why
ΔXZW ≅ ΔXZY

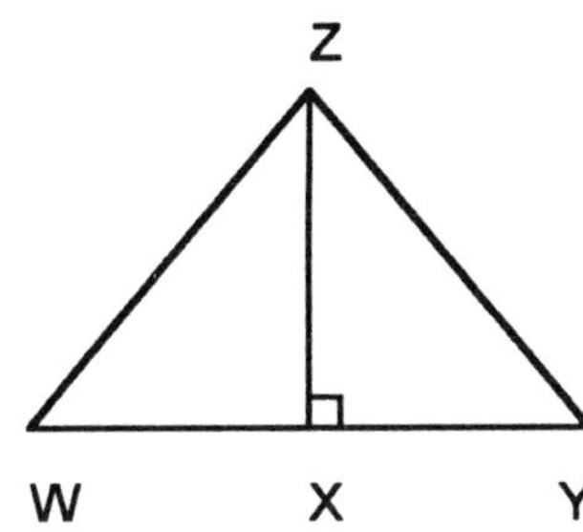

11. M bisects PN and LO
Show why PL = ON

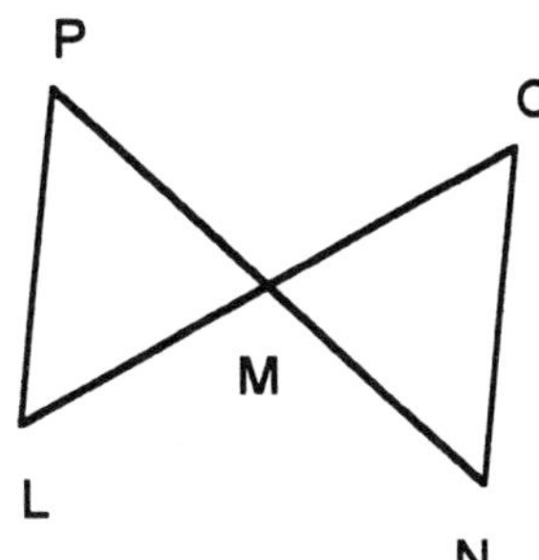

12. ∠D = 90°, AC = 10
ΔADC is isosceles
Show ΔABD ≅ ΔCBD

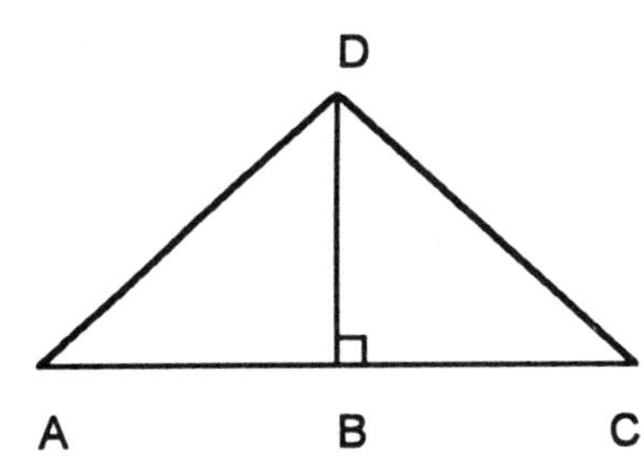

13. Find x

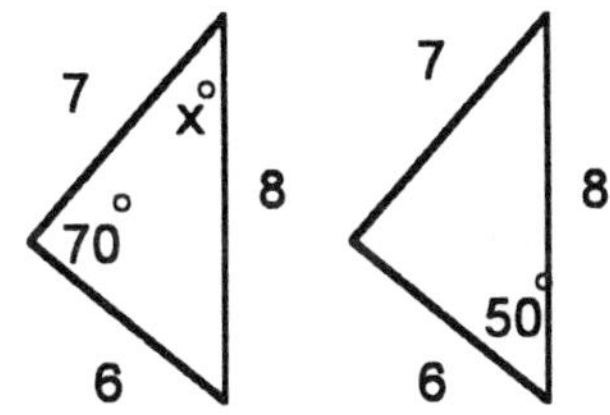

14. Find x

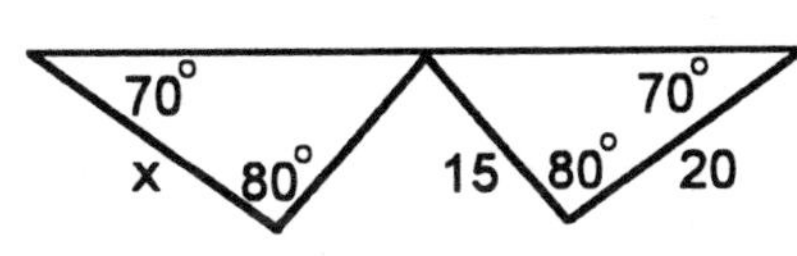

15. Find x

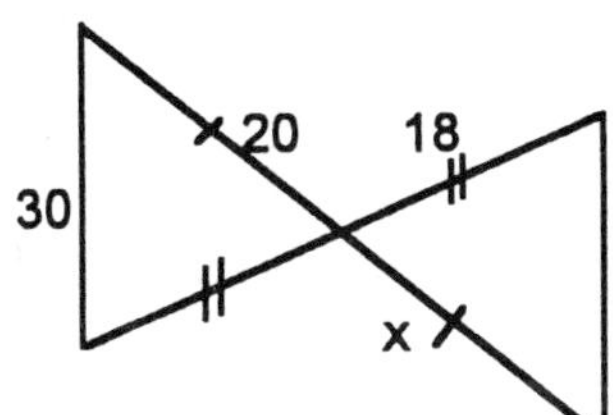

Answers to ⌘ problems: Chapter 3

⌘ 1 a) vertex D corresponds to vertex W, so $\angle D = \angle \mathbf{W}$

AR corresponds to ZT, and FD corresponds to FW, so $\frac{AR}{FD} = \frac{ZT}{\mathbf{FW}}$

⌘ 2 a) $\frac{16 \text{ dimes}}{5 \text{ quarters}} = \frac{160 \text{ cents}}{125 \text{ cents}} = \mathbf{\frac{32}{25}}$

b) $\frac{144}{24} = \mathbf{\frac{6}{1}}$

c) $\frac{130 \text{ square feet}}{26 \text{ gallons}} = \frac{5 \text{ square feet}}{1 \text{ gallon}} =$

5 square feet per gallon

d) $\mathbf{\frac{3}{5} = \frac{9}{15}}$

e) $\frac{3}{4} = \frac{x}{24}$

cross-multiply $4x = 72$

solve $\mathbf{x = 18}$

f) Methods I and II

25 shirts cost \$640

how many shirts cost \$192

set it up $\frac{25 \text{ shirts}}{\$640 \text{ cost}} = \frac{x \text{ shirts}}{\$192 \text{ cost}}$ $\qquad \frac{25 \text{ shirts}}{x} = \frac{640}{192}$

cross-multiply $640x = 4800$ $\qquad 640x = 4800$

solve $\mathbf{x = 7.5}$ $\qquad \mathbf{x = 7.5}$

Assuming that you could only get as many shirts as you could completely pay for, you could get **7 shirts.**

g) Solve: there are 40 - 22 = 18 men in a class of 40 students, so:

18 men in a class of 40

how many men in a class of 140

set-up $\frac{18 \text{ men}}{40 \text{ std}} = \frac{x \text{ men}}{140 \text{ std}}$ $\qquad \frac{18 \text{ men}}{x \text{ men}} = \frac{40 \text{ std}}{140 \text{ std}}$

cross-multiply $40x = 2520$ $\qquad 40x = 2520$

solve **x = 63 men** $\qquad$ **x = 63 men**

⌘3 a) if you work out the two missing angles (20 and 30), determining the corresponding vertices is straightforward. The proportion becomes:

$$\frac{20}{x} = \frac{24}{y} = \frac{30}{18}$$

which yields $\frac{20}{x} = \frac{30}{18}$ $\qquad \frac{24}{y} = \frac{30}{18}$

cross-multiplying $30x = 360$ $\qquad 30y = 432$

solving $\mathbf{x = 12}$ $\qquad \mathbf{y = 14.4}$

b) Noting that the parallel lines will make the diagonally opposite angles equal, the proportion sets up as:

$$\frac{x}{15} = \frac{8}{y} = \frac{10}{25}$$

which yields	$\frac{x}{15} = \frac{10}{25}$	$\frac{8}{y} = \frac{10}{25}$
cross-multiplying	$25x = 150$	$10y = 200$
solving	$\mathbf{x = 6}$	$\mathbf{y = 20}$

⌘ **4** a)
CE = EB
AE = ED
$\angle CED = \angle AEB$
$\Delta ECD \cong \Delta AEB$
SAS $\cong$ SAS

b)
GK = GF
GF = JH
$\Delta FKG \cong \Delta JKH$
HL $\cong$ HL

c)
LM = MN
PM = PM
$\Delta NMP \cong \Delta LMP$
SAS $\cong$ SAS

or

LM = MN
PM $\perp$ LN therefore, ΔLNP is isosceles
LP = PN
$\Delta NMP \cong \Delta LMP$
HL $\cong$ HL

d) Look at small triangle ΔQRS(i), pull apart 2 triangles ΔUQR and ΔTRQ(ii)

i) $\angle SQR = \angle SRQ$ (isos.) QR = QR ii) $\angle UQR = \angle TRQ$ (right angles)

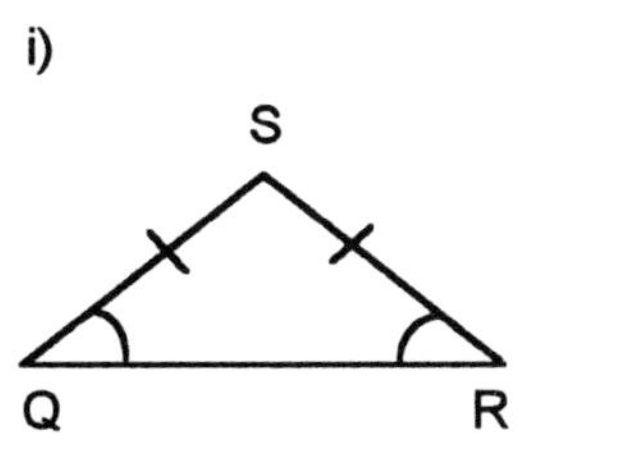

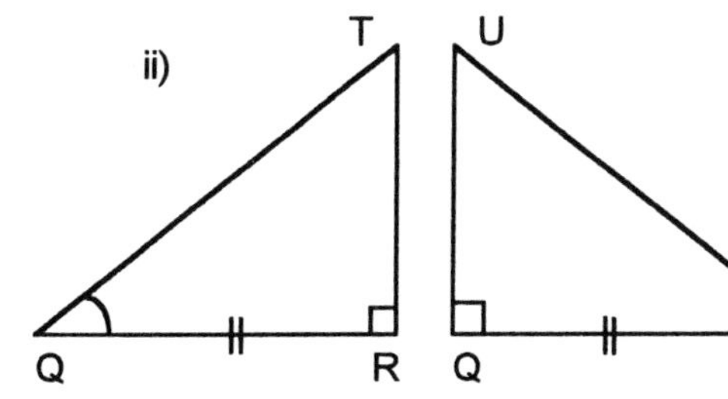

$\Delta NMP \cong \Delta LMP$, ASA $\cong$ ASA

⌘ **5** a) The vertical line is an altitude, bisecting the base. x = 90°, y = 90 - 40 = 50°, and $z = \frac{36}{2} = 18$ The solution is **x = 90°, y = 50°, z = 18**

b) The altitude to the base bisects the vertex angle, so x + 80 = 90, x = 10. y is found by the Pythagorean theorem:

$$y^2 + 36^2 = 39^2$$

$$y^2 + 1296 = 1521$$

$$y^2 = 225$$

$$\mathbf{y = 15}$$

PART 2
LOGIC

CHAPTER 4 LOGIC I: THE BASICS

Topics in this chapter:

- Statements: Their Validity and Truth Values
- Quantifiers and Venn Diagrams
- Compound Statements (Using And and Or)
- Negation
- Arguments: Their Validity and Truth Values

Answers to ⌘ try-out exercises are given at end of the chapter.
(3b) means that this item or idea is illustrated in **Figure 3, part b.**

4.1 Logic statements, validity, and truth values

The basic building block of logic is the *statement*. **A valid logic statement is a sentence which is either TRUE or FALSE.** There are a lot of sentences in the world, but only some of them will qualify as a logic statement. The sentence cannot be a question, wish, or guess. It cannot be an unsure situation or "maybe" or "sometimes". **The validity of the statement has <u>nothing</u> to do with whether its truth value is TRUE or FALSE.**

<u>Examples:</u>	<u>validity and truth value, if any</u>
1) 5 > 3.	**valid statement,** TRUE (math)
2) 5 - 2 = 8	**valid statement,** FALSE (math)
3) x is even.	**not a valid statement** - truth cannot be determined unless x has a value
4) I am happy.	**valid statement** - truth value of this opinion has to be determined by person asserting it.
5) He is handsome	**not a valid a statement** - truth cannot be determined until "he" has a value,
6) It is raining.	**valid statement** - truth can be determined
7) Is it raining?	**not a valid statement,** even though the answer (yes/no) can be determined (question)
8) It will rain tomorrow	**valid** - "for the sake of argument" in a "what-if" argument (some will not permit this)
9) I wish it would rain	**not a valid statement** (wish)
10) Ants are tiny	**valid statement** - truth value depends on a judgment call - how small is "tiny"?

- **The critical thing for determining whether or not a sentence is a valid statement is:**
  could you determine whether or not the sentence is TRUE or FALSE?
  In general, if it is a question, an exclamation, a wish, or lacks a specific subject, it is not a valid statement.
- This book will accept opinions, guesses, and predictions as valid statements, provided they have specific subjects and can be declared TRUE or FALSE "for the sake of argument". This is not too off-the-wall, when you consider that much of what is presented as fact today is a "best guess" or a wish. Some books, texts, teachers, and tests may not wish to accept these as valid statements.
- Even though whether it is TRUE or FALSE does not matter for determining its validity as a logic statement, the TRUE/FALSE value will matter later in logic arguments in determining the truth value of the argument (but not its validity)

⌘ 1 Are the following *valid statements* in the logic sense?

a) Mr. Garcia, my history teacher, has brown hair.
b) I wonder if it will rain tomorrow.
c) I think last night's algebra homework was hard.
d) $5 - 2 > 1$

4.2 Quantifiers and Venn diagrams

The Venn diagram is a very useful visual method of presenting logic statements. This method was developed by John Venn and Leonhard Euler. These Venn-Euler diagrams (or Venn diagrams for short) are informal and easy to draw. A statement is represented by dots for individuals and circles, ovals, ellipses, or even irregularly-shaped blobs for groups or conditions. We will label all of the group-representations "circles." All points inside a circle marked "A" have that quality "A", and all the points outside the circle do not have the quality "A".
(1) shows the Venn diagrams for two simple statements.
In **(1a),** the circle represents math majors, with all points inside it representing "is a math major", and all points outside of it representing "is not a math major."
In **(1b),** if Juan is represented by a dot, then "Juan is a math major" is shown by putting the dot inside the circle.
In **(1c),** "Robert is not a math major" is a dot outside the circle.

Figure 1 - elementary Venn diagrams

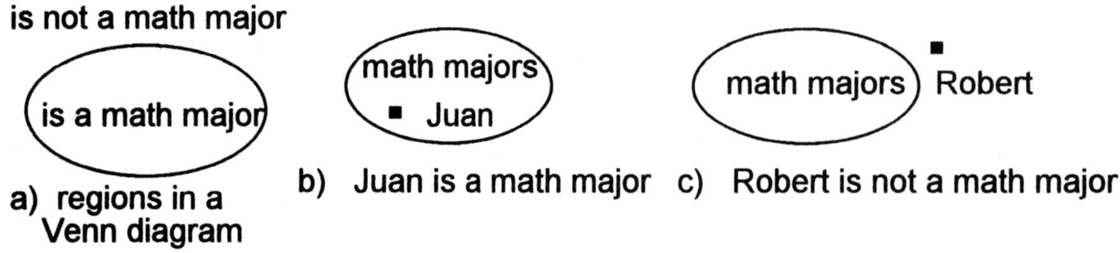

a) regions in a Venn diagram b) Juan is a math major c) Robert is not a math major

Quantifiers are used to modify a statement by changing its universality. There are three families of quantifiers. Here are some of the quantifiers in each family:

all, every, each; **some,** at least one, not all; **no,** none of, not one

If the original statement is *This apple is red.*, the same statement with a quantifier added might be *No apples are red.*, *Some apples are red.*, or *All apples are red.* Be careful about hidden assumptions: *Apples are red.* is many times informally understood to mean *All apples are red.*

The truth value of a statement can change with the addition of a quantifier.

If you declare that	*All apples are red.*	is TRUE,
this means that the equivalents	*Every apple is red.*	and
	Each apple is red.	are TRUE.
In contrast,	*No apples are red.*	
	Not all apples are red.	
	Some apples are red.	and
	At least one apple is red.	are FALSE.

The reason that the first of these statements is FALSE is obvious, but the last three take a bit of thought to see the reason. The last three are FALSE because they deny that every apple is red or suggest that all apples being red is only a possibility and not a certainty..

Quantifiers, by their very nature, refer more to groups of items rather than individual items. For example, "All horses are gray" refers to more than one horse. Groups (all horses) as well as conditions (are gray) are represented in Venn diagrams by circles. **(2)** shows several examples Venn diagrams for statements with quantifiers.

In **(2a),** the horse circle is inside the gray circle, because we wish every point in the horse circle to be in the gray circle for "All horses are gray.".

In **(2b),** there is an overlap between the gray circle and the horse circle, representing horses which are gray. The overlap is not complete, so that some horses are gray, and some are not. Also, there are some gray things which are not horses. The letter **a** is a horse which is not gray, **b** is a gray horse, **c** is gray but not a horse, and **d** is neither a horse nor is it gray.

In **(2c),** there is no overlap, so no horse points are in the gray circle, representing "No horses are gray." Also, then, no gray things are horses. The overlap is an extremely important concept called the **intersection** (a set-theory term).

Be wary of informal (and logically sloppy) use: **all** means "absolutely every one", rather than "most" or "almost all"; and **none** means "not a single one" rather than "just a few."

Figure 2 - quantifiers in Venn diagrams

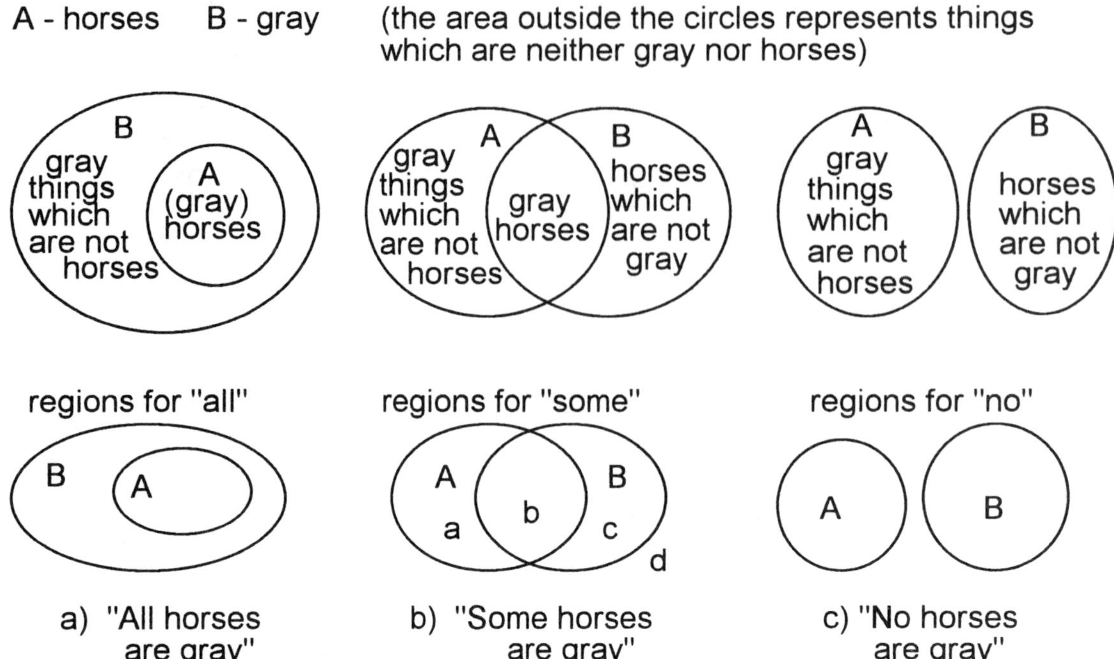

⌘**2** Give a statement for each of the Venn diagrams in **Drawing 1.** (the exact words in the answers may vary from person to person.)

Drawing 1 **A - convertible** **B - red cars** **C - air conditioned cars**

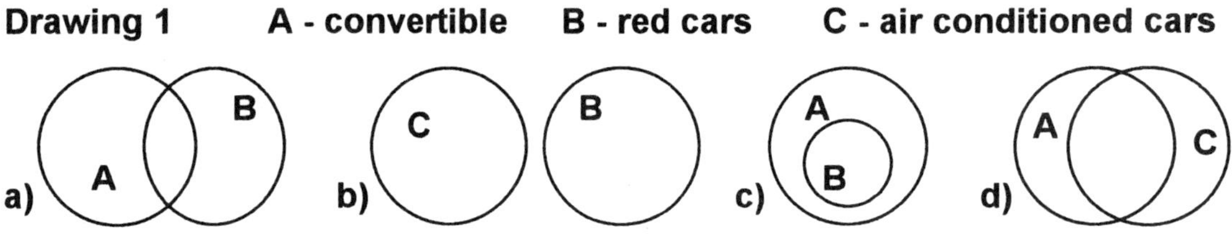

⌘ **3** Give a Venn diagram for each of these statements:

a) Marci is a gymnast.
b) Fernando does not like video games.
c) Some people who eat at Mexican Restaurants like their food spiced extra hot and some demand fast service. Margarita likes extra hot food served quickly when she eats at a Mexican Restaurant

⌘ **4** Express the Venn diagrams in **Drawing 2** as sentences (answers may vary.)

Drawing 2

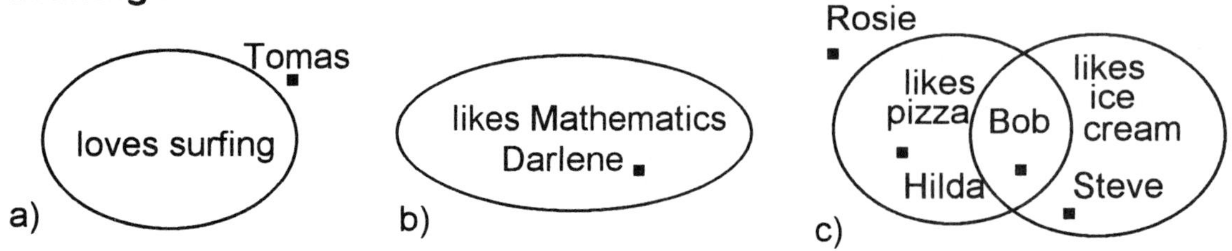

4.3 Compound Statements (and, or)

Compound statements are made by putting two simple logic statements together with **and** or **or.** If **A** and **B** are valid statements, the two types are:

two simple statements connected by **and** (**A and B**)
two simple statements connected by **or** (**A or B**)

Examples: suppose the simple statements are

A: *This rose is red*
B: *Austin is the capital of Texas*

Result: the two kinds of compound statements would be:

*This rose is red, **and** Austin is the capital of Texas* (**A and B**)
*This rose is red, **or** Austin is the capital of Texas* (**A or B**)

Statements made with **and** are sometimes called **conjunctions,** and represent the **intersection** (every item in the overlap) in set theory. Those put together with **or** are sometimes called **disjunctions,** and represent the **union** (every thing in both circles) in set theory.

A compound statement's truth value depends on the truth values of its two simple statements:.

A compound statement made with **and** is TRUE only if both of the simple statements are TRUE. If either or both of the simple statements is FALSE, the compound statement is FALSE. Thus, *2 is an even number **and** 3 is a multiple of 5* is FALSE because the *3 is a multiple of 5* is FALSE.

A compound statement made with **or** is TRUE if either one or both of the simple statements are TRUE. For example, *Blue is a primary color **or** Washington, DC is on the West Coast* is TRUE, even though the simple statement *Washington, DC is on the West Coast* is FALSE.

With either **and** or **or,** if both simple statements are FALSE, the compound statement is FALSE. Thus, both *Green is orange **and** 5 is an even number* and *Flies are mammals* **or** *water is dry* are FALSE compound statements.

Many people prefer to summarize compounding truth value facts with these tables:

TRUE **AND** TRUE = TRUE	TRUE **OR** TRUE = TRUE
TRUE **AND** FALSE = FALSE	TRUE **OR** FALSE = TRUE
FALSE **AND** TRUE = FALSE	FALSE **OR** TRUE = TRUE
FALSE **AND** FALSE = FALSE	FALSE **OR** FALSE = FALSE

⌘ **5** Are the following statements true or false?

a) France is in Europe or Germany is in South America.
b) Violets are fire-engine red-orange or roses are electric blue.
c) The Oboe is a percussion instrument and Austin is the capitol of Texas.

EXERCISES 4.1 - 4.3

Are the following *valid statements* in the logic sense?

1. Milk is a good source of calcium.
2. He rides the bus to work every working day.
3. x is an even number.
4. Ice cream melts if it is left out in the sun too long.
5. I wonder who will win the world series this year.
6. $8 > 4$
7. Is 4 + 6 equal to 10?
8. Is a yard bigger than a meter?
9. $7 - 3 = 2$
10. $9 + 1 = 11$
11. I wish the price of good concert tickets would come down.
12. The Dodgers are going to win their next game.

Write a sentence describing each situation portrayed by a Venn diagram in **Drawing 1.**

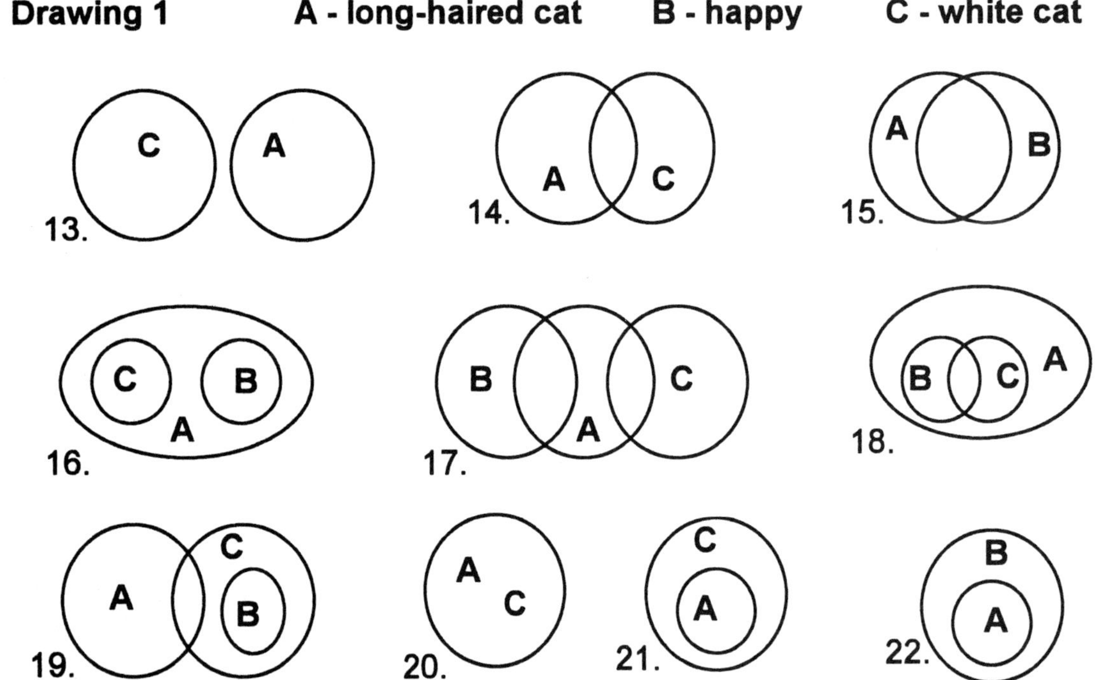

Write a sentence for the Venn diagrams in **Drawing 2.** Answers may vary.
Three names are given and represented by: ♣ , ♦ , ♥

Drawing 2 A - long-haired, B - dog Wiggles: ♣ Duke: ♦ Midnight: □♥

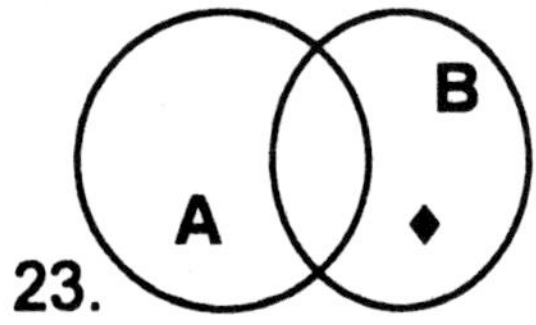

24. A B □♥

25. A □♥ B

26. A B ♣

Express the following statements with Venn diagrams:

27. Jeremy is a basketball player.
28. Francesca is not a sun-lover.
29. All students enjoy spring break.
30. All dogs like bones.
31. No student enjoys flunking a test.
32. No athlete plays a sport without warming up.
33. Some students take college art.
34. Some girls liked Elvis, but Margie is not one of them.

Are the following compound statements true or false?

35. Chili is a spicy dish or California is an Eastern state.
36. Math is a symbolic language and most birds can fly.
37. Triangles have three sides and squares have five sides.
38. Gray is a neutral color or electric blue is not a bright color.
39. $6 > 10$ or $9 < 11$
40. Water is wet or dust is dry.
41. 10 pennies make a dime and 5 nickels make a half dollar.
42. Six plus nine is fourteen and seven is less than ten.

4.4 Negation

Negation of a simple statement
Negation can be a tricky subject when done in English. We had something of the same effect when we applied the quantifier <u>all</u> to *This rose is red* and got *<u>All</u> roses are red* - a change from singular to plural. If we wish to negate *Roses are red*, we would need to apply a **not** to the sentence.
Here are the guidelines for where to put the **not:**

> **If the sentence has a being verb such as *is*, put the not on the predicate's noun or adjective.** Thus the negation of *Roses are red* is *Roses are **not** red.*
>
> **Other kinds of verbs: put the <u>not</u> on the verb.**
> *Brad has a hat* negates to *Brad does **not** have a hat.* The statement *Brad has no hat* looks just as good but is not equally correct because the *<u>no</u>* is a quantifier.

Here are some examples of the negation of a simple statement:

statement	negation of the statement
Angela has a book.	*Angela does not have a book.*
Candy does not like Opera.	*Candy likes Opera.*
Maria is an A student.	*Maria is not an A student.*

Negation always reverses the truth value of a statement. It makes a true statement false, or a false statement true. In the Venn diagram **(3), not** moves the point from the inside of the circle to the outside, or from outside the circle to inside the circle.

Figure 3 - negation of simple statements

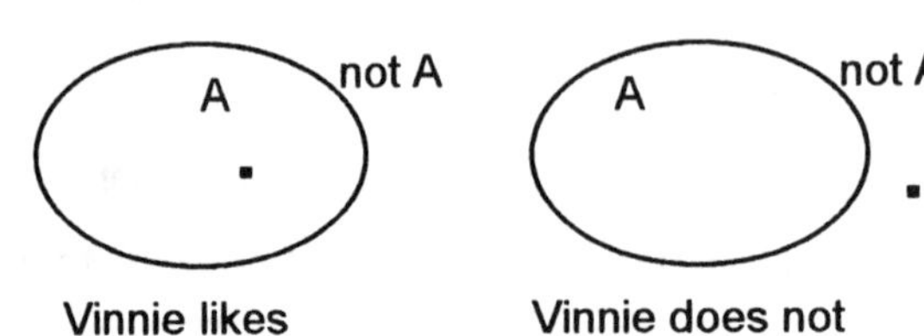

Vinnie likes fast cars

Vinnie does not like fast cars

Negation of quantifiers

The negation of quantifiers must be done most carefully. **(Figure 4)**

Figure 4 - negation of quantifiers

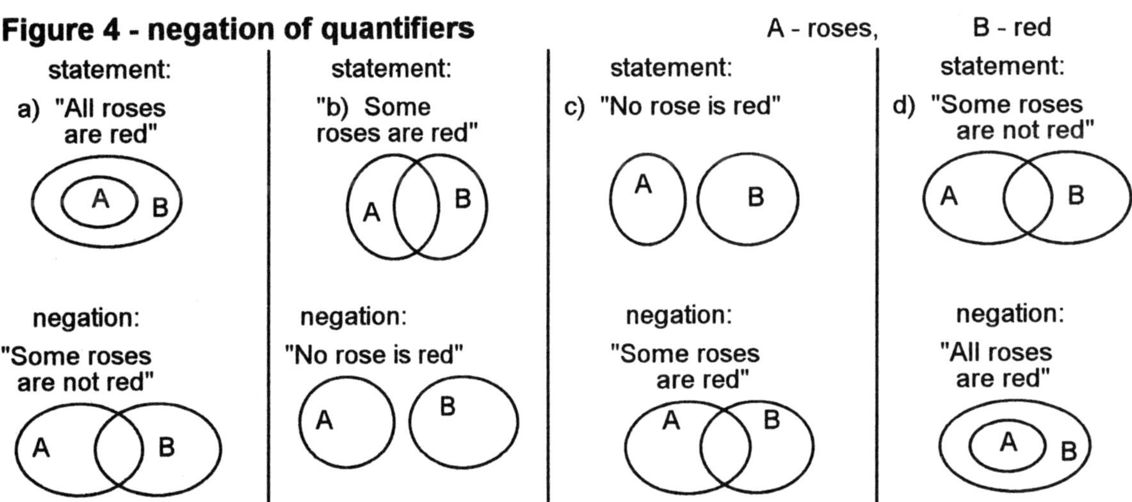

Details - (optional): Why the negation of *All roses are red* is *Some roses are not red*. Let us look at this in detail. *All roses are red* means every single rose is red. If you had 100 roses at the start, all 100 would be red.

Then negation could mean that:

1 rose is not red, and 99 are red
or 49 roses are not red and 51 are red
or 100 roses are not red and 0 are red.

So, between 1 and 100 of the roses are not red. At least one is still red. This is best expressed by *Some roses are **not** red.*

The following table summarizes these facts:

statement	negation
all A are B	some A are not B
some A are B	no A are B
no A are B	some A are B
some A are not B	all A are B

⌘ 6 Negate the following statements:

a) Some students are full-time students.
b) Not all roses are red. (same as Some roses are not red)
c) No citizens like taxes.
d) All sea gulls are independent.
e) Baseball players do not like rain delays.

Negation of compound statements

There are special rules for negating compound statements called **DeMorgan's Laws:**

The negation of A **or** B is **not** A **and not** B
The negation of A **and** B is **not** A **or not** B

Example:
The **negation** of *Texas is hot in the summer **and** the north pole is always cold* is *Texas is **not** hot in the summer **or** the north pole is **not** always cold.*

As you might expect, **if you negate any kind of statement you reverse its truth value..** In this case the original statement was TRUE, and the negation is FALSE.

⌘ 7 Negate these statements:

a) Fish swim and birds fly.
b) Julio likes soccer or Jana does not like rap music.

4.5 Logic arguments and their validity

A logic argument is a sequence of two or more statements (called premises) and a result (called the conclusion). The validity of an argument depends on the validity of the reasoning used to reach the conclusion, and not on the truth values of the statements.. A valid argument can yield a truth value of FALSE.

Here is an example of an argument:

Argument #1

All math majors like science.	premise
Edward Carlos is a math major.	premise
therefore, Edward Carlos likes science.	conclusion

This is a valid argument. Even though the truth value of *All math majors like science* is debatable, **all statements are treated as TRUE for the determination of an argument's validity.** (This is where the phrase "accept it for the sake of argument" comes from.) The best way to examine the validity of an argument is with Venn diagrams.

Figure 6 - argument #1 and Venn diagrams

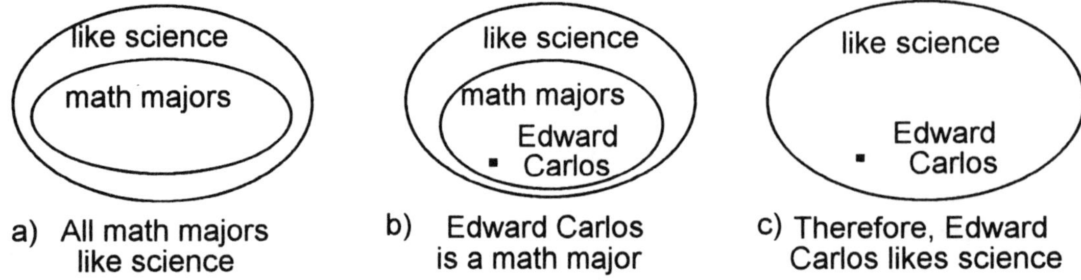

a) All math majors like science

b) Edward Carlos is a math major

c) Therefore, Edward Carlos likes science

In **(6a),** the Venn diagram for *All math majors like science* is displayed.

In **(6b)** is the Venn diagram showing *Edward Carlos is a math* major.

In **(6c),** because *Edward Carlos'* dot is in the *math major* circle, it is also in the *likes science* circle. Thus the conclusion: *Edward Carlos likes science.*
This is a valid argument.

Here is another example of an argument:

Argument #2

Some roses are pink flowers.
Cecilia gave me a pink flower.

---

therefore, Cecilia gave me a rose.

Setting up the Venn diagrams for argument #2:

Figure 7 - Venn diagrams for argument #2 a flower ♣

roses | pink flowers

"Some roses are pink"

roses ♣ pink flowers

roses pink♣ flowers

"a pink flower" -♣

there are two possible locations for♣

" a (pink) rose" (the overlap) is not the ONLY possibility, so it cannot be the valid conclusion

As you can see, this particular argument was not valid because there were two possible ways to place the marker for *a pink flower* - one was in the part of the *rose* circle which was the overlap, which would have made it *a pink rose*. However, since you have no reason in the argument to throw away the other possibility which would put the marker in *the pink non-rose* area, you could not claim that the ONLY possibility was *a pink rose*. **If you can come up with more than one possible end situation (solution), the argument is not valid,** therefore it is called a **fallacy.**

Sometimes you are given the task of taking a couple of statements and deriving your own conclusion. For example, what would be a valid conclusion for:

Everyone who eats breakfast at Munch'N'Go spends too much.
Roger ate breakfast at Munch'N'Go this morning.

Using Venn diagrams, it becomes fairly obvious that if the ♥ marker is in the *eating breakfast at Munch'N'Go* circle, it is also in the *spending too much* circle. Therefore a valid conclusion would be *Roger spent too much.*

Figure 8 - steps in coming to a conclusion

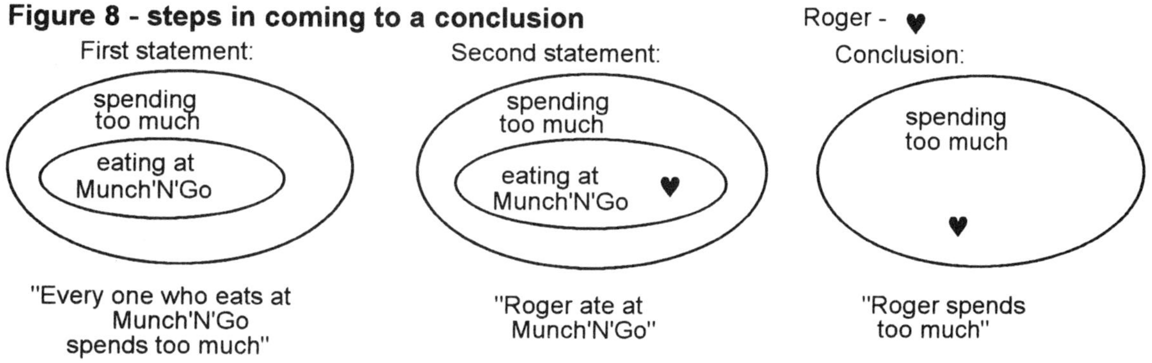

Roger - ♥

First statement:

spending too much
eating at Munch'N'Go

"Every one who eats at Munch'N'Go spends too much"

Second statement:

spending too much
eating at Munch'N'Go ♥

"Roger ate at Munch'N'Go"

Conclusion:

spending too much
♥

"Roger spends too much"

⌘ **8** Are these arguments valid?

a) Nancy is a superb cook.
Superb cooks use a lot of dishes.
____
Nancy always uses a lot of dishes.

b) Gilbert likes fishing.
People who like to dream fish a lot.
____
Gilbert dreams a lot.

c) Some oak trees lose their leaves in the fall.
The tree outside my window is an oak tree.
____
The tree outside my window will lose its leaves in the fall.

⌘ **9** Write a Valid conclusion for each argument, or put "no conclusion"

a) Anyone who likes computers is good at using them.
Larry is good at using a computer.
____

b) Cindy goes bowling a lot.
Bowling makes you happy,.
____

c) Some dogs are black.
Some cats are black.
Reginald has black fur.
____

Implications

Implications are often called "if-then" statements. They are really not that far from the kinds of statements we have been working with thus far. For example, consider this sequence:

All apples are red	(quantifier)
If it is an apple, it's red	
If it is an apple, then it is red.	(if-then)

The implication *if **A**, then **B**,* where ***A*** and ***B*** are statements, has names for its parts. ***A*** is called the **hypothesis,** and ***B*** is called the **conclusion.**
The implication *if A, then B* can also be worded as

A** implies **B
A**, only if **B
B** follows from **A
B** is necessary for **A
A** is sufficient for **B
not-A**, or **B

An implication ***A** implies **B*** is true as long as **B** is true, regardless of **A**'s truth value.

There are three important valid statements related to the original statement.

Original	If A then B	If you have black hair, then you are a brunette.
Converse	If B then A	If you are a brunette, then you have black hair.
Inverse	If not-A, then not-B	If you do not have black hair, then you are not a brunette.
Contrapositive	If not-B, then not-A	If you are not a brunette, then you do not have black hair.

The Contrapositive is always TRUE if the original statement is TRUE
The Inverse is always TRUE if the Converse is TRUE.
BUT - The Converse is not necessarily TRUE, even if the original statement is TRUE.

We will deal mainly with the statement and its converse in this book, and only lightly with the inverse and contrapositive. The common mistake of trying to use the converse in an argument is that generally it does not lead to a valid argument. For example, the statement "If you are from Chicago, then you like pizza" has the converse "If you like pizza, then you are from Chicago." Even if the original statement is TRUE, which is debatable, the converse certainly cannot be claimed to be TRUE - surely other people in the world besides Chicagoans like pizza.

The converse statement is always as valid as the original statement, but is not necessarily TRUE. Claiming the converse is TRUE is a common mistake. It is the source of many socio-political statements whose truth cannot be asserted. Inappropriate ethnic, religious, political or gender references are common blunders of converse statements not always being true.

The example about brunettes and black hair is a **definition.** The converse of a definition is always TRUE if the definition is TRUE. Many definitions and other fundamental statements are explicit about this - they will say "If A is TRUE, then B is TRUE, and conversely." Definitions and statements whose converse is TRUE are often stated as biconditionals or two-way implications "A implies B and B implies A": Another way to say this is "A is TRUE if and only if B is TRUE." Mathematicians have shortened "**if and only if**" to "**iff**". The other common way of saying a definition or biconditonal statement is "**A is necessary and sufficient for B.**

The negation of an implication : the negation of ***A*** *implies* ***B*** is ***A*** *and not* ***B***. This is reasonable if you use DeMorgan's Laws to negate the last of the alternative statements for ***A*** *implies* ***B***.

⌘ **10** Write the converse, inverse, and contrapositive for the following statement. If the original statement is assumed to be true, what are the truth values of the converse, inverse, and contrapositive?
statement: Dinosaurs are not alive today.

EXERCISES 4.4 - 4.5

Negate the following statements:

1. Vicente is a good athlete.
2. Kelly has a lot of homework to grade.
3. Ellen does not tolerate lazy students.
4. Bob is not afraid of computers.
5. Nelly never takes tests on time. (Nelly does not take tests on time.)
6. Bernado has never been a football fan.
7. Not all candy is made with sugar
8. Leroy always watches his favorite soccer team on TV.
9. All cooks have a lot of cookbooks.
10. Some fishermen always use live bait.

11. No hiker ever passes up a nice day for a hike.
12. Some calculus problems are not easy.
13. Some frozen food is low-calorie.
14. All bats are nocturnal.
15. Some chairs do not have a back.
16. At least one of the clowns was sad.
17. Each day has 24 hours.

Negate the following compound statements:

18. Purple is not a primary color, or gray is a neutral color.
19. Flowers need sun, and flowers do not need insect pests.
20. Fermin is a teacher and math requires a lot of study.
21. Angela does not like geometry or Lionel does not like algebra.
22. Leandro collects Navajo pottery or Inez likes to collect Alaskan artifacts

Are the following arguments valid?

23. Anyone who likes plants will have many plants.
Jill likes plants.

Therefore, Jill has many plants.

24. If you like western music, then you will like this show.
But since you don't like this show,

You do not like western music.

25. An animal that chases cats is mean.
This dog does not chase cats.

Therefore, this dog is not mean.

26. Older houses do not cost much.
This house does not cost much.

Therefore, this is an older house.

27. All rectangles have parallel sides.
This polygon has parallel sides.

This polygon is a rectangle.

28. Porcupines are spiny.
This animal is not spiny.

This animal is not a porcupine.

29. If you do not study, the test will be much harder.
You find the test easy.

Therefore, you studied for the test.

30. All black surfaces get very hot in the sunlight.
This black surface is in the sunlight.

This surface will get very hot.

31. If Nalo is not elected, taxes will to up.
Nalo was not elected.

Therefore, taxes will go up.

32. If an animal is not cold-blooded, it is a mammal.
This animal is not a mammal.

Therefore, it is cold-blooded.

33. If you practice a lot,
you can be a Karate Champ.
Mike is a Karate Champ.

Mike practices a lot.

34. Julie likes music very much.
People who do not like music
never smile

Julie always smiles

35. All trees are green.
This chair is green.

Therefore this chair is a tree.

36. Some dogs like cats.
All cats eat birds.

Therefore some dogs eat birds.

37. Some summer nights are cloudy.
If it is cloudy, you cannot
see the stars.
Tonight, you cannot see the stars.

Therefore, it is summer.

38. No logicians like ice cream.
Some people who like ice cream
study law.
Philip does not like ice cream..

Therefore, Philip is a logician.

Write a valid conclusion to the argument:

39. Black cats are friendly.
Whiskers is a friendly cat.

____

40. Armadillos live in Texas.
Texas is in the U.S.A.

____

41. If I study, I will pass the test.
I will study.

____

42. If you like cookies, you will like this.
Jason likes cookies.

____

43. Red roses smell good.
This is a white rose.

____

44. Some roses are red.
This is a rose.

____

Write the converse, inverse, and contrapositive for the following statements.

45. statement: If Roy does not have a free weekend, he will not visit his grandkids.
46. statement: If it is not free, it will cost something.
47. statement: A rectangle with four equal sides is a square.
48. statement: If it is a flower, then it is pretty.
49. If I have to go, then you have to go too.
50. If it is raining, then we can not go fishing.

Answers and some reasons to ⌘ problems: Chapter 4

⌘ 1 a) **valid statement** - truth can be determined

b) **not** a valid statement

c) **valid statement** - at least by our rules

d) **valid statement** - true

⌘ 2 These are some of the possible valid answers.

a) Some red cars are convertibles. (or Some convertibles are red.)

b) No red cars have air conditioning.

c) All red cars are convertibles.

d) Some convertibles are air conditioned.

⌘ 3

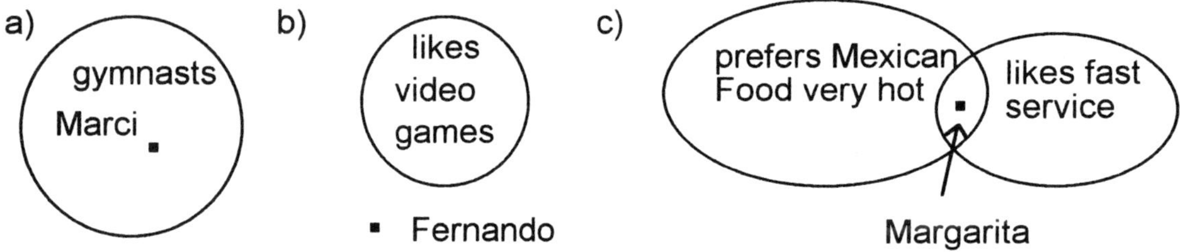

⌘ 4 These are some of the possible valid answers.

a) Tomas does not love surfing.

b) Darlene likes Mathematics

c) Hilda likes pizza, Steve likes ice cream, Bob likes both pizza and ice cream, and Rosie likes neither ice cream nor pizza.

⌘ 5

a) France is in Europe - TRUE
Germany is in South America - FALSE
TRUE or FALSE is **TRUE**

b) Violets are fire-engine red-orange - FALSE
roses are electric blue - FALSE
FALSE or FALSE is **FALSE**

c) The Oboe is a percussion instrument - FALSE
Austin is the capitol of Texas - TRUE
FALSE and TRUE is **FALSE**

⌘ 6 answers may vary

a) No students are full-time students.

b) All roses are red.

c) Some citizens like taxes.

d) Some sea gulls are not independent.

e) Some baseball players like rain delays.

⌘ 7 a) Fish do not swim or birds do not fly.

b) Julio does not like soccer and Jana likes rap music.

⌘ 8

a)

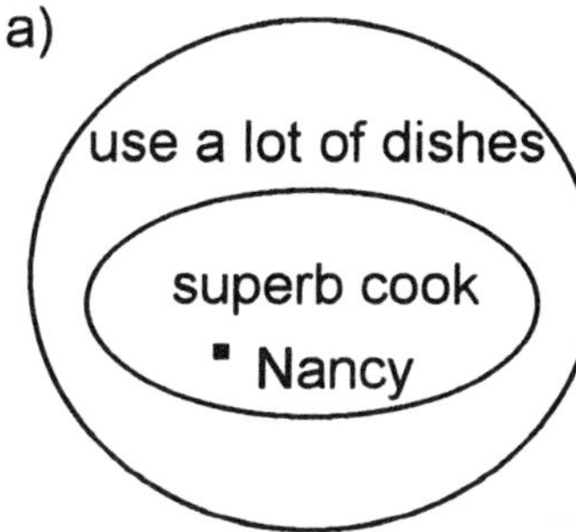

Nancy is in the "uses a lot of dishes" circle, so the argument is valid

b)

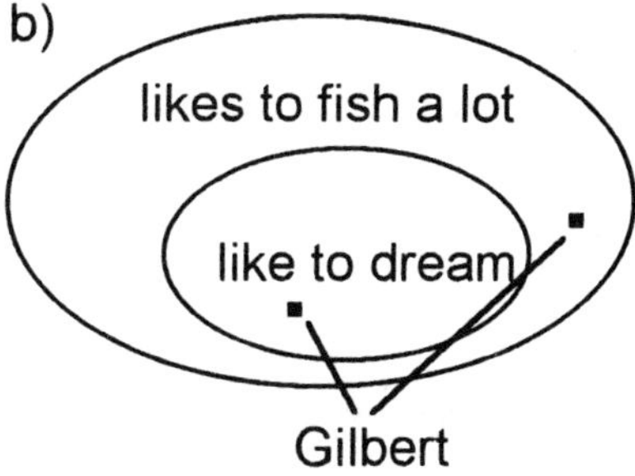

Gilbert's dot in the "likes to fish a lot" circle could land in two different spots, only one of which is in the "likes to dream circle," so it is not a valid argument.

c)

The dot for the oak tree outside my window is in two circles, so it is not a valid argument.

⌘ 9

a)

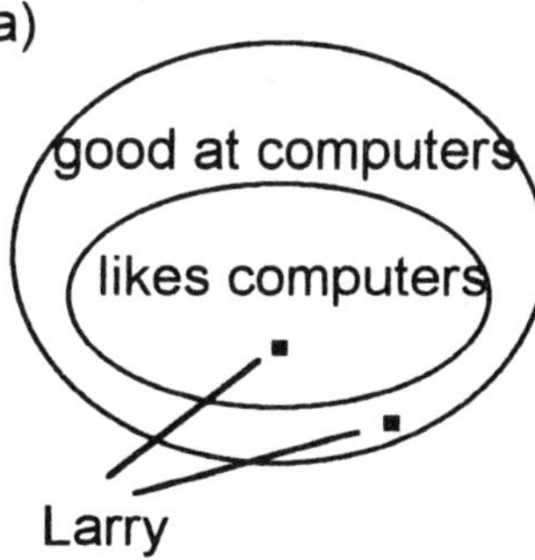

Larry's dot is in two different places, so there is no valid conclusion.

b)

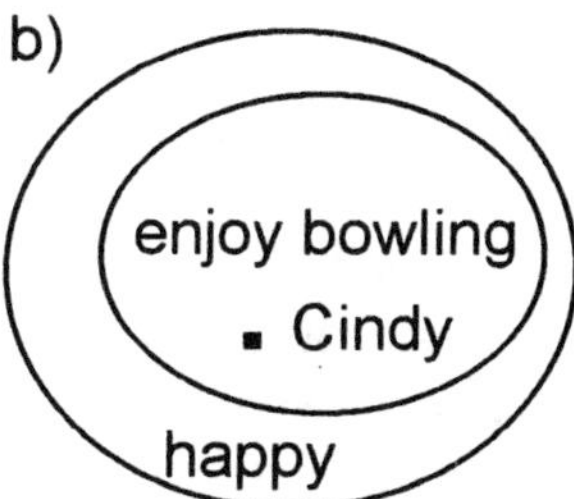

Since Cindy's dot is in "enjoy bowling" and "enjoy bowling" is in the "happy" circle, Cindy's dot is in the "happy" circle, which implies that Cindy is happy.

c)

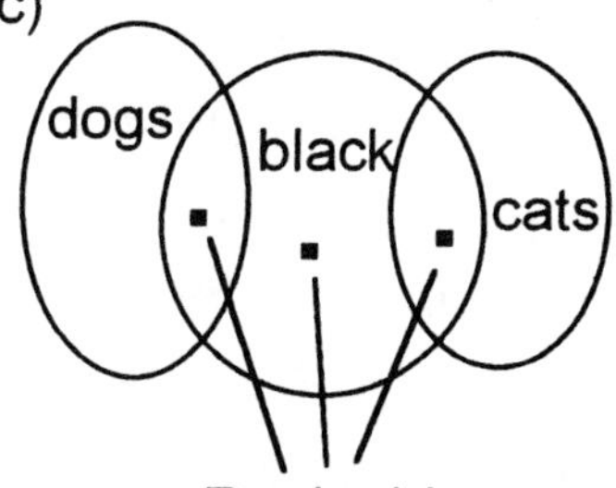

Reginald's dot is in three different places, so there is no valid conclusion.

⌘ 10

statement: Dinosaurs are not alive today. (assumed true)

converse: If it is not alive today, it is a dinosaur..(false)

inverse: If it is not a dinosaur, it is alive today. (false)

contrapositive: If it is alive today, it is not a dinosaur. (true)

CHAPTER 5 LOGIC 2: REASONING

Topics in this chapter:
- ♦ Inductive reasoning
- ♦ Deductive reasoning
- ♦ An application: Sequence problem

Answers to ⌘ try-out exercises are given at end of the chapter.
(3b) means that this item or idea is illustrated in **Figure 3, part b.**

5.1 Inductive Reasoning

Inductive reasoning is the process of looking at several specific cases and discovering a pattern. It is getting to a general end result - a pattern - based on certain particular observations. **The key to inductive reasoning is discovering the pattern.** For example, suppose that you notice that every time that the sky has heavy clouds in it and the wind shifts around to the north, it rains. Therefore you say "There is a pattern: heavy clouds and the wind out of the north means that it will rain". You have just drawn a pattern out from life. Let's see some other examples:

> You received a letter from Senator Smith in the mail on Monday. Your friend Carlos got a letter from Senator Smith on Monday too. Your Aunt Maria also got a letter from Senator Smith on Monday. You see a pattern: "Most people got a letter from Senator Smith last Monday".
>
> You counted the raisins in 4 boxes of your favorite cereal, and each time you came up with 239 raisins. Therefore you induced, or discovered, the pattern: "There are 239 raisins in a box of your favorite cereal".
>
> The last 15 cookies you bought at the mall were stale. Therefore you conclude: "Cookies bought at the mall are stale "

In each example, several similar situations were examined, each having a similar outcome, leading to the discovery of a pattern. In business, industry, and engineering, these rules are often called "rules of thumb."

There are **two cautions** which need to be put in here:

1) **Avoid faulty generalizations by making too broad an assertion**
   Example: You buy two dozen <u>cantaloupes</u> from XY-Mart and find that they are still green, and assert: "<u>All produce</u> sold by XY-Marts is bad."
2) **Avoid "jumping to conclusions" from insufficient data**
   Example: you get poor service <u>once</u> at a restaurant, and assert: "That place <u>always</u> has lousy service!"

These half-serious examples are instances where, if human beings are involved, lead to all kinds of inter-personal, inter-racial, and political misunderstandings and hatreds.

Many inductive rules are true. Most are at least "probably true". A few are false. In each case, we have discovered a general rule or pattern by examining a set of specific or particular situations. The conclusion drawn may be correct or incorrect (depending on the correctness of the input data), but in each case the reasoning process was valid, as long as sufficient observations are made and the urge to over-generalize is avoided.. Again: **validity and truthfulness are not the same thing.**

Try to find the rule or pattern in these sequences:

⌘1 The state lottery has a "pick 6 to win" game. The first week, one of the winning numbers was 34. The next week, one of the numbers is 34. Last week, 34 was one of the numbers.
Conclusion (rule): ________________________________

⌘2 If you double 4, you'll get 8, an even number. If you double 17, you'll get 34, an even number. If you double -10, you'll get -20, an even number. If you double 0, you'll get 0, an even number.
Conclusion (rule): ________________________________

5.2 Deductive Reasoning

Deductive reasoning takes a pattern or rule and applies the rule to a specific case to get the end result. For example, if the pattern is "all sports cars are red", and your friend buys a sports car, using deductive reasoning you can assert "my friend's car is red". Here are some other examples of this kind of deductive reasoning.

Rule: all lemons are sour.
Therefore, this particular lemon must be sour.

Pattern: if I don't study for a test, I flunk it.
Therefore, if I don't study for the test on Friday, I will flunk it.

Rule: all quadrilaterals have 4 sides.
So, if someone tells me a figure is a quadrilateral, I know that it has 4 sides.

Pattern: if I stay up late, I will want to sleep in the next morning.
So, if I stay up late on Wednesday night, I will want to sleep in on Thursday morning.

Try to reach a conclusion from these patterns:

⌘ 3 Pattern: all oranges have seeds.
This is an orange. Therefore, ________________

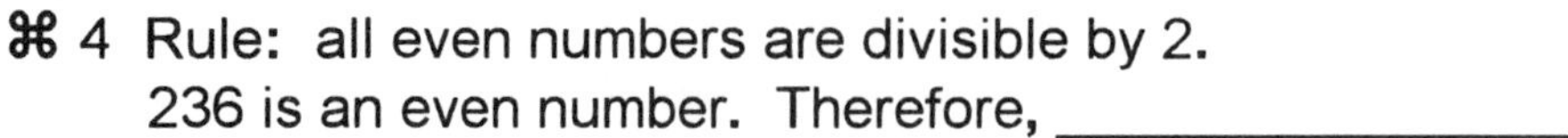

⌘ 4 Rule: all even numbers are divisible by 2.
236 is an even number. Therefore, ________________

In general, rules are stronger than patterns. Most rules come out of highly structured areas of knowledge, such as Mathematics or Grammar. These rules are usually accepted as valid with no argument. However, a "rule of thumb" developed out of patterns noticed in everyday experience can be every bit as valid.

In reality, the separation of the thinking process into inductive and deductive parts is somewhat artificial, because in everyday life we usually induce-deduce as a pair of actions. The only places that much deductive reasoning is used on its own are in science (applying formulas) and mathematics (constructing proofs and applying formulas). To get those theorems and formulas in the first place some scientist or mathematician had to notice a pattern in order to get a formula to derive or a theorem to prove. In science, after a pattern is deduced it is confirmed by successfully predicting some further event for verification of the pattern. This is sometimes called "The Hypothetico-deductive method" or "The Scientific Method".

We considered the validity of deductive logical arguments in Chapter 4. We also learned that the only circumstances under which a deduction (implication) is false was if the outcome is false. For the "sake of argument" in this chapter, treat all assertions as true. We will put deduction to work on some puzzles in Chapter 6.

The next topic we will consider is a special kind of induction-deduction process which is used to find the next member of a sequence. Several standardized tests these days have sequence problems in them like:

- What is the next number? 2, 4, 6, 8, __

The pattern or rule might be "counting by 2's", or "add 2 each time to the number before" (**GET THE PATTERN - INDUCTION**). Using that as the rule, the next number is 8 + 2 = 10. (**APPLY THE PATTERN - DEDUCTION**)

5.3 Sequence Problems - Applying Induction and Deduction

In mathematical number sequences, the pattern (inductive) will help us to find the next number (deductive). Here we will consider some methods of working on the more difficult problems. Sequences are formed by starting with some number, letter, or pattern, and applying some rule repeatedly to that starting item and to its results. We will first consider number sequences. Example: if we start with the number 15, and use the rule "add 3 each time", the first result would be 15 + 3 or 18, the next result would be 18 + 3 or 21, the next result would be 21 + 3 or 24, and so on. This would yield the sequence:

15, 18, 21, 24, and so on

- What would be the next few numbers in that sequence?

The end product would be the sequence 15, 18, 21, 24, 27, 30, 33, 36, . . .   The . . . at the end means "it keeps going on forever, using the same rule to get from each number to the next". It is official math-language for "and so on" or "etc.,"

The sequence problems usually start out by giving several members of the sequence and then asking you to supply the next one or several members of the sequence. For example, our example above would usually be presented as

What is the next number in   15, 18, 21, 24, 27, ___ ?

For a sequence problem, your objective is to:

#1 look at the sequence to find the rule (inductive part)

#2 apply the rule to get the next member of the sequence (deductive part)

The real challenge is usually #1. Fortunately, there are normally a limited number of things that go into the rule:

Adding or subtracting the same number
Multiplying or dividing by the same number
Taking each member of some sequence and raising each member to the same power
Combinations of the 3 rules above

In addition, there are unique or oddball sequences which follow unusual rules.

Let's take the sequence we have used for an example and see if we can work backwards (the inductive part) to find the rule. Then we can apply the rule and get the next member of the sequence - the answer (the deductive part).
When you first encounter the sequence the most helpful thing to do is to 1) write it down, leaving some space between the numbers, and some space below them. Then 2) find the difference between each pair of successive numbers. 3) Repeat this process until the difference are the same. Then 4) apply the rule to the last number to get the answer.

Write it down

15   18   21   24   27

Take the difference between each number and the next, and write them on the line below the numbers, forming the "first level".

15   18   21   24   27
   3   3   3   3

Notice that the difference is the same each time. When the difference is the same each time, your search is over! Now you apply the rule to the last number (27 + 3) and that is the next number. **Your answer is 30.**

⌘ 5. Try it out on 13, 18, 23, 28, 33, __

Sometimes the sequence can require several subtraction-steps to get to the rule. For example, let's see if we can get the rule for 1, 7, 20, 40, 67, __

Again, write out the numbers, leaving some space:

1 7 20 40 67

Next take the differences

	1		7		20		40		67
level 1:		6		13		20		27	

Then take the differences again (the "level 2" differences)

	1			7			20			40		67
level 1:		6			13			20			27	
level 2:			7			7			7			

The "second level" differences are the same: 7. Using this pattern, 27 + 7 gives the next "first level" difference of 34. Then the next member of the sequence is 67 + 34, or 101.

	1			7			20			40			67			101
level 1:		6			13			20			27			34		
level 2:			7			7			7			7				

⌘ 6. Try this: what is the next number in: 10, 13, 24, 43, 70, ___

Sometimes you may have to go down to a third level of differences. Let's check this out in a slightly different type of problem: 146, 108, 81, 62, 48, ___.

Notice that previously the sequences increased from number to number, but this sequence decreases from number to number. We will still take differences to figure out the rule, but then we will subtract to get the answer. As before, write down the numbers with some spaces and subtract.

146 108 81 62 48

Get the "first level" differences:

	146		108		81		62		48
level 1:		38		27		19		14	

Then get the "second level" differences:

	146			108			81			62		48
level 1:		38			27			19			14	
level 2:			11			8			5			

And again to get the "third level" differences:

	146		108		81		62		48
level 1:		38		27		19		14	
level 2:			11		8		5		
level 3:				3		3			

The third level differences are the same. Then work backwards - the next "third level" difference is the same - 3; the next "second level" difference is 5 - 3 or 2, and then the next "first level" difference is 14 - 2 or 12, so the next number in the sequence is 48 - 12 or 36 (the answer at last!).

	146		108		81		62		48		36
level 1:		38		27		19		14		12	
level 2:			11		8		5		2		
level 3:				3		3		3			

⌘ 7. Try this: 47, 36, 27, 20, 15, __

The kind of problems that we have been working on is known as an arithmetic sequence, because the same thing is added or subtracted from each successive number to get the next number in the sequence. If each successive number is multiplied or divided by the same number, it is called a geometric sequence. An example of a geometric sequence is: 2, 6, 18, 54, 162.

- Can you discover the rule is for the sequence 2, 6, 18, 54, 162?

If you guessed "multiply each number by 3", you were right! But let's see how you can get the answer without guessing. To work this sort of problem out, again, write out the numbers with some spaces:

2 6 18 54 162

This time, divide each number in the sequence by the number in front of it.

	2		6		18		54		162
level 1:		3		3		3		3	yielding a constant value of 3.

Thus the next number will be 3 x 164 or 492 (the answer). This was an increasing sequence, with each successive number getting larger. If the sequence numbers had gotten smaller instead, you would have divided each number by the one after it. Sequences using a multiplication or division rule rarely go beyond the first level.

⌘ 8. Try this: 12, 3, $\frac{3}{4}$, $\frac{3}{16}$, $\frac{3}{64}$, ___

In addition to all of the rules so far, the members of the sequence may alternate in sign (the same as being multiplied by a -1), such as: $\frac{2}{3}$, -2, 6, -18, 54, where, starting with the $\frac{2}{3}$, each successive number is multiplied by -3.

Another kind of sequence is goes "back and forth", or adds something, subtracts something, adds something, subtracts something, etc.

For example, take a good look at: 17, 13, 19, 11, 21, 9, 23, . . . The rule here is: start with 17, take away 4, add 6, take away 8, add 10, take away 12, add 14, . . . (difference each time of 2). Another way you can think about this kind of sequence is by separating the increasing part and decreasing parts out like this:

	17		19		21		23
17 13 19 11 21 9 23 splits out to							
		13		11		9	

Then the pattern is more obvious. In this case, the differences always changed by a difference of 2 - the top sequence adds 2 and the bottom sequence subtracts 2. The next two numbers in the sequence are 7 and 25. The amount added or subtracted might be different, but the true nature of the sequence is usually obvious when you split out the two sub-sequences. If you come upon a back-and-forth sequence, your first step should be to find differences, just as before, and then take a look at what is happening.

⌘ 9. Try this: 65, 74, 62, 77, 59, __

Another form of sequence is one which uses powers of numbers. The simple example is: 4,9,16, 25, 36, ___.

If you know your squares, this one should be easy to spot. (The next number is the next perfect square, 49.) Sometimes these are used in conjunction with other things, yielding sequences such as:

1, 8, 27, 64, ___	(sequence of cubes, with the next being 125.)
63, 31, 15, 7, 3, 1, __	(sequence of powers of 2, minus 1. The next one would be 2^1 - 1, or 0)
2, 8, 18, 32, 50, ___	(2 times the sequence of squares, 1, 4, 9, 16, 25, with the next number 2 x 36 or 72, or this one can be worked as an arithmetic sequence if you take it to "level 2").
17, 18, 21, 26, 33, __	(17+0, 17+1, 17+4, 17+9, 17+16, with the next one being 17+25, or 42, or it can be worked as a "level 2" sequence)
1, 4, 16, 64, 256 ___	(The even powers of 2 (2^0, 2^2, 2^4, 2^6, 2^8), or powers of 4, or 4x1, 4 times that result, etc.)

As you can see, there is often more than one rule which will fit these sequences, or more than one way of thinking out the solution. This is a general rule for any sequence for which only a limited number of numbers are given.

Alphabetic sequences are usually like the arithmetic sequences, but with numbers associated with the letters. Which number usually depends on the particular example, such as:

Find the next letter: a, d, g, j, m, ___ .

Here you can associate "a" with 1, "d" with 4 (d is the 4th letter of the alphabet), "g" with 7 (g is the 7th letter of the alphabet), "j" with 10, and "m" with 13. The number difference is a constant 3, so the next letter has the number 13+3, or 16. Thus the next letter is the 16th letter, "p". Rather than trying to make a formal rule, it is usually easier to think something like:

"a, b-skip, c-skip, d,
d, e-skip, f-skip, g,
g, h-skip, i-skip, j,
j, k-skip, l-skip, m - - - I did 2 skips on each of those, so if I do 2 more after m...
m, n-skip, o-skip, p! the answer!"

The majority of alphabetical sequences can be easily solved this way. If the problem looks complex, or involves going backwards through the alphabet, you might want to write down the alphabet (or as much of it as is necessary) and look for a pattern or assign numbers and calculate differences. The main difference with alphabetical problems is that there is a first or lowest letter, a, and a last or highest letter, z. If your rule takes you past them you will have to make a decision on how to handle it. Use the alphabet listed on the next page to try the examples below:

A B C D E F G H I J K L M N O P Q R S T U V W X Y Z

⌘ 10. Find the next letter: z, x, v, t, r, ___

⌘ 11. Find the next letter: f, g, i, l, p, ___

⌘ 12. Find the next letter: a, g, m, s, y, ___

The category of unique and oddball sequences: these are not used by standardized placement or achievement tests, but sneaky puzzle books and some math teachers have been known to use them. Each one of these sequences has a rule. Just for fun, see if you can find the rule and next item for these sequences. Remember, except for ⌘ 15., the rules aren't like any we have discussed so far.

⌘ 13. Try this: O, T, T, F, F, S, S, E,__

⌘ 14. Try this: 10000, 01000, 00100, 00010, __

⌘ 15. Try this: 1000, 1001, 1010, 1011, 1100, __

⌘ 16. Try this: S, S, M, T, W, T, __

⌘ 17. Try this: 3, 33, 4, 44, 5, __

⌘ 18. Try this: 1, 22, 333, 4444, __

⌘ 19. Try this: 1, 1, 2, 3, 5, 8, 13, 21, 34, 55, __

Another type of sequence is made up of drawings. The usual thing happening here is that something rotates or changes in a sequence, or maybe two things. Figure out what changes, and then choose the next member.

Examples:

- If the top 4 drawings are the first four in a sequence, then which of the bottom drawings would be the next one?

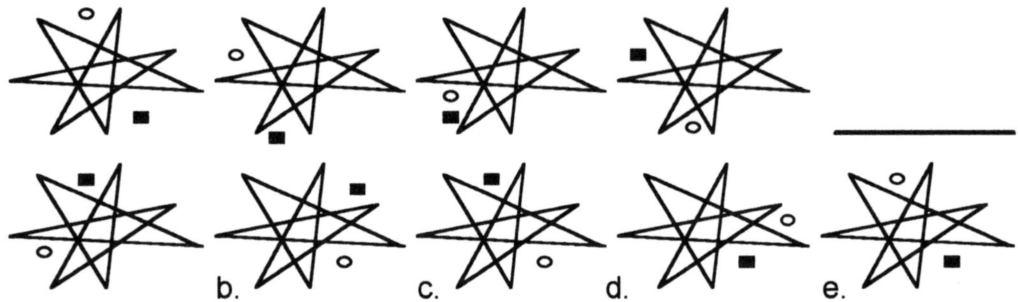

There are two things changing around a constant star: a white circle and a black square. The white circle is rotating one position (notch between spikes) counter-clockwise (to the left on the top), and the black square is rotating one position clockwise (to the left on the bottom). The next position should be pattern c.

- Which drawing of the bottom 5 should be next after the top 4?

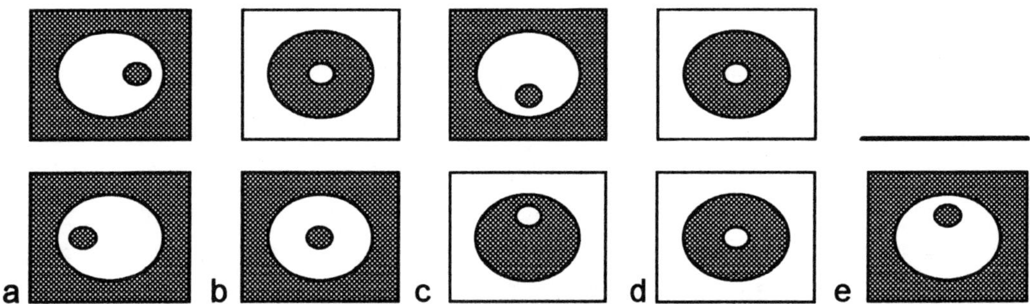

The first and third patterns change (the little circle rotates 1/4 of the circle in a clockwise direction) and the second and fourth patterns do not change. Therefore, the fifth pattern will change by having the little circle rotate one more 1/4 of the circle clockwise, yielding pattern a.

CHAPTER 5 EXERCISES

A B C D E F G H I J K L M N O P Q R S T U V W X Y Z

Find the next item in each sequence:

1. 2, 5, 8, 11, 14, ___
2. 27, 23, 19, 15, 11, ___
3. 17, 33, 56, 86, ___
4. 7, 8, 11, 16, 23, ___
5. 16, 24, 30, 34, 36, ___
6. 67, 60, 54, 49, 45,___
7. y, v, s, p, m, ___
8. A, H, N, R, U, ___
9. 1, 5, 14, 30, 55, ___
10. 15, 12, 8, 3, -3, ___
11. 9, 5, 8, 6, 7, ___
12. 4, 9, 5, 8, 6, 7, ___
13. M, L, N, K, O, J, ___
14. N, K, P, H, T, ___
15. 2, 10, 30, 68, 130, ___
16. 5, 8, 13, 20, 29, ___
17. 2, 3, 5, 8, 13, ___
18. $1, \sqrt{2}, \sqrt{3}, 2, \sqrt{5},$ ___
19. $\frac{1}{2}, \frac{2}{3}, \frac{3}{4}, \frac{4}{5},$ ___
20 $2, 1, \frac{2}{3}, \frac{1}{2}, \frac{2}{5},$ ___
21. $3, \frac{3}{4}, \frac{1}{3}, \frac{3}{16},$ ___
22. XIX, VXIXV, MVXIXVM, XMVXIXVMX, VXMVXIXVMXV, ___

From the second row, choose the best answer to be the next item in the first row

23.

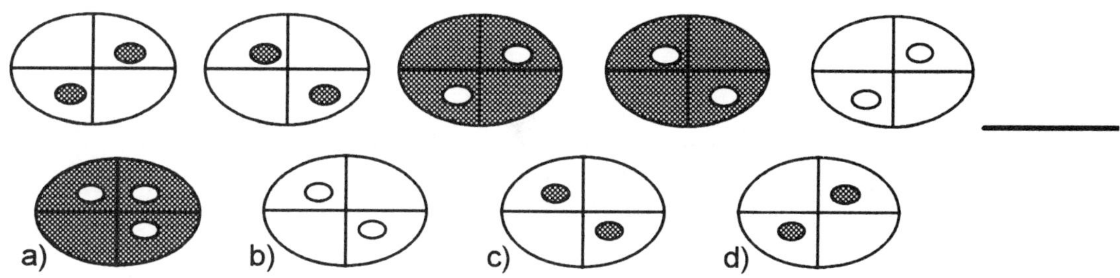

24.

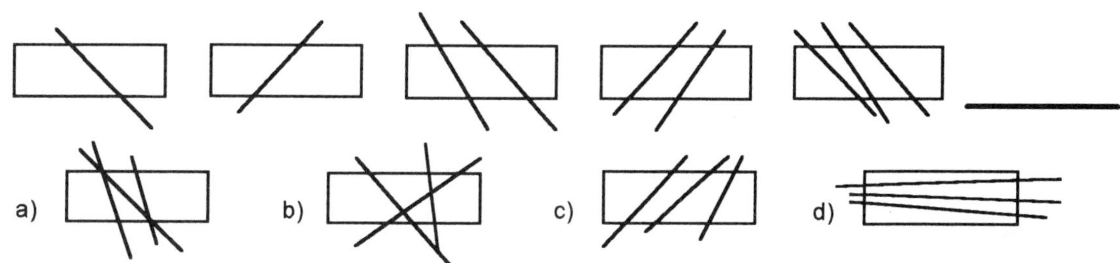

25.

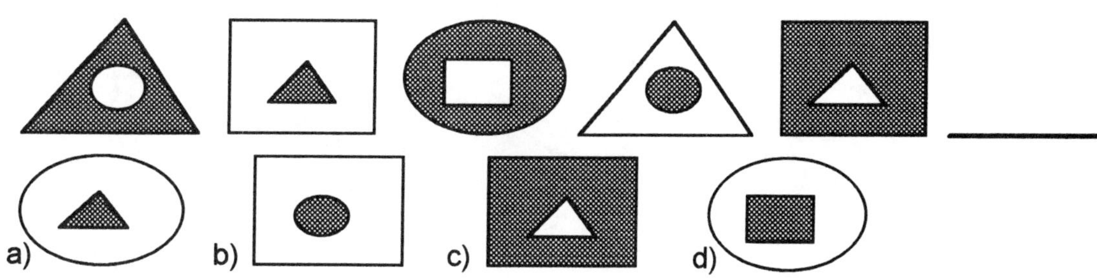

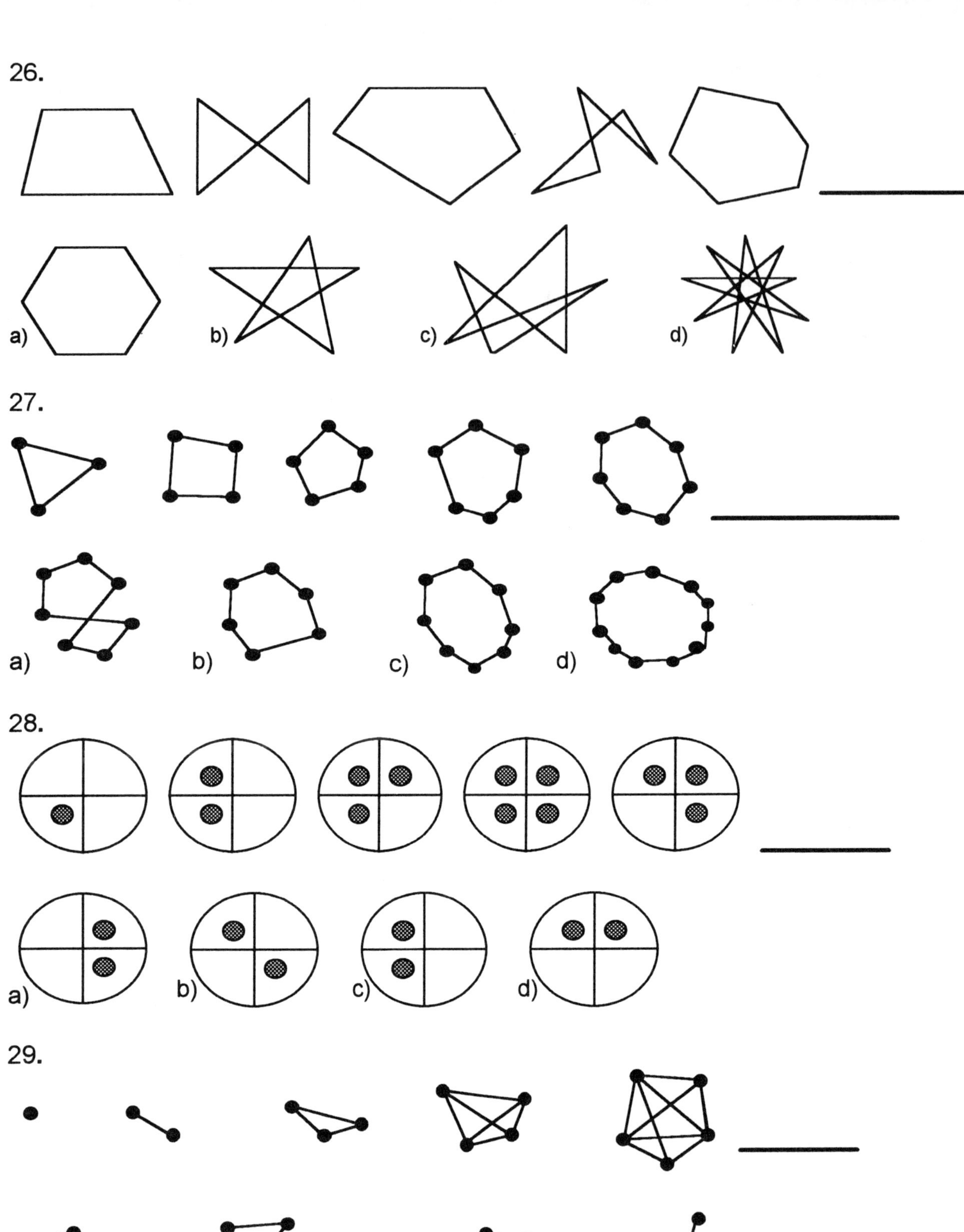
26.
a)
b)
c)
d)
27.
a)
b)
c)
d)
28.
a)
b)
c)
d)
29.
a)
b)
c)
d)

Answers and some reasons to ⌘ problems: Chapter 5

⌘ 1 The rule seems to be that 34 is picked every time, leading to the rule: **if there is a "pick 6 to win" game, 34 will be one of the numbers picked.**

⌘ 2. No matter what kind of integer you start with (negative, positive, zero, odd, or even), when you double it you get an even number of some sort. This leads to the rule: **any integer will yield an even number when doubled.**

⌘ 3. **Therefore, this (orange) has seeds.** ⌘ 4. **Therefore, 236 is divisible by 2.**

⌘ 5. 18 13 23 28 33

5 5 5 5 The difference each time is 5, so we find the next number by adding 5 to the last one given: 33 + 5 = 38. **The answer is 38.**

⌘ 6. 10 13 24 43 70

3 11 19 27

8 8 8 The difference on level 2 each time is 8, so we find the next number by adding 8 to the last level 1 difference: 27 + 8 = 35. Then we add this next level 1 difference to the last number given, 70: 70 + 35 = 105. **The answer is 105.**

⌘ 7. 47 36 27 20 15

11 9 7 5

2 2 2 The difference on level 2 each time is 2, so we find the next level 1 number by subtracting 2 from the last level 1 number: 5 - 2 = 3. This 3 is then subtracted from the last number given: 15 - 3 = 12. **The answer is 12.**

⌘ 8. This looks like a multiplication-type sequence. Divide each number by the number in front of it, and by the time you get down to level 1, the ratio each time is $\frac{1}{4}$

$12 \quad 3 \quad \frac{3}{4} \quad \frac{3}{16} \quad \frac{3}{64}$

$\frac{3}{12} \quad \frac{\frac{3}{4}}{3} \quad \frac{\frac{3}{16}}{\frac{3}{4}} \quad \frac{\frac{3}{64}}{\frac{3}{16}}$

or $\frac{1}{4} \quad \frac{1}{4} \quad \frac{1}{4} \quad \frac{1}{4}$ The next number is then the last given number times $\frac{1}{4}$, or $\frac{3}{64} \times \frac{1}{4} = \frac{3}{256}$. **The answer is $\frac{3}{256}$.**

⌘ 9. 65 74 62 77 59 splits out to

65 62 59

74 77

and the top row decreases by 3 each time, and the bottom row increases by 3. The next number on the bottom row will be 77 + 3 = 80. **The answer is 80.**

⌘ 10. z, x, v, t, r Go backwards through the alphabet. It may help to list the alphabet first, starting at about m because the sequence deals with the last part of it: m n o p q r s t u v w x y z. Then, working backward, z, skip-y, x, skip-w, v, skip-u, t, skip-s, r, skip-q, p. **The answer is p.**

⌘ 11. f, g, i, l, p List the alphabet and mark the letters given.
a b c d e **f** **g** h **i** j k **l** m n o **p** q r s t u v w x y z.
The pattern should be fairly obvious now: Start at **f**, skip none, **g**, skip 1, **i**, skip 2, **l**, skip 3, **p**. So we should start with p and skip 4, which gives u.
The answer is u.

⌘ 12. a, g, m, s, y List the alphabet and mark the letters given.
a b c d e f **g** h i j k l **m** n o p q r **s** t u v w x **y** z. The pattern is: to mark every sixth letter starting with a, or skip 5 each time. To get the answer you must decide what to do - you will run out of alphabet. Most likely the best solution is to just wrap-around and start at a again after z. Write out the continuation and skip 5 after y:
a b c d e f **g** h i j k l **m** n o p q r **s** t u v w x **y** z a b c d **e** f g
The answer is e.

⌘ 13. One Two Three Four Five Six Seven Eight . . . **Nine,** so **the answer is n**

⌘ 14. 10000, 01000, 00100, 00010, . . .This is more of a visual pattern and does not at all depend on numerical values - it is just a "1" migrating to the right through a bunch of zeroes. The next place puts the "1" after the last 0, so **the answer is 00001.**

⌘ 15. 1000, 1001, 1010, 1011, 1100, . . . This is binary (base 2) counting for 8, 9, 10, 11, 12. The next one will be binary for 13: 1101. **The answer is 1101.**

⌘ 16. Saturday, Sunday, Monday, Tuesday, Wednesday, Thursday, . . . Friday,
The answer is F

⌘ 17. 3, 33, 4, 44, 5, . . . This is another pattern sequence. **The answer is 55.**

⌘ 18. 1, 22, 333, 4444, . . ; This is also a pattern sequence. **The answer is 55555.**

⌘ 19. 1, 1, 2, 3, 5, 8, 13, 21, 34, 55, . . . This is a historically famous sequence called the Fibonacci Sequence. Every member of the sequence after the first two is the sum of the previous TWO numbers. So, the next number after 55 would be 55 + 34 = 89. **The answer is 89.**

CHAPTER 6 LOGIC 3: MORE APPLICATIONS

Topics in this chapter:

- ♦ Survey (Counting) Problems
- ♦ Order Problems
- ♦ Deductive Puzzle Problems

Answers to ⌘ try-out exercises are given at end of the chapter.
(3b) means that this item or idea is illustrated in **Figure 3, part b.**

6.1 Survey (Counting) Problems

Survey problems are so named because they usually start out "A survey was done on a group of 500 car owners, and it found that . . .". Let us pursue this problem and discover the principles involved and common questions asked.

"A survey was done on a group of 500 car owners, and it found that 170 of them owned white cars, 330 of them owned cars with a cassette deck, and 100 owned white cars with a cassette deck. How many owned a car which was not white and did not have a cassette deck?" A helpful way to approach this problem is by using a Venn diagram.

Figure 1 - Venn diagram for a counting problem

a) in general (the box represents the entire survey group)

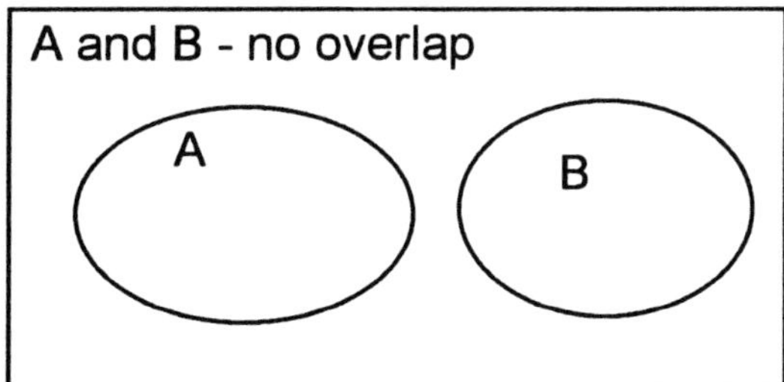

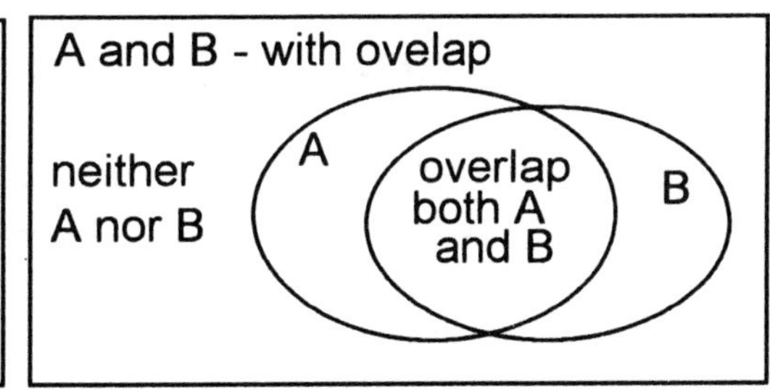

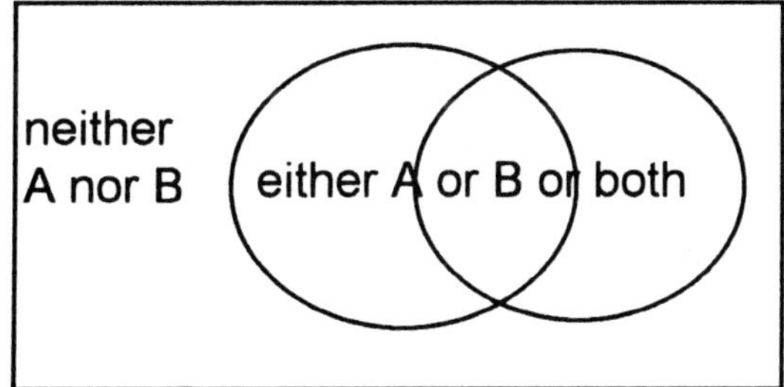

b) for this specific problem
500 car owners

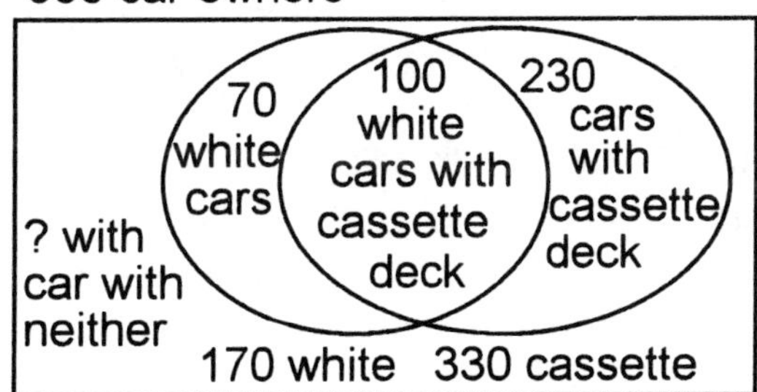

The rules, or formulas, which relate to counting problems are:

Rule 1: The number which are in either group A or group B or both = the number in the A group + the number in the B group - the number which are simultaneously in both the A group and the B group (the overlap).

Either A or B or both = A + B - both A and B (overlap)

Rule 2: The number in the entire survey = the number in neither the A group nor in the B group + the number in either group A or group B or both.

Everyone surveyed = either A or B or both + neither A nor B

Rule 1 looks strange until you consider the Venn diagrams in **(1a)** and **(2).** Part of the **A** circle is the overlap, and part of **B** is the overlap. When you add the **A** circle items to the **B** circle items, the overlap gets added in twice. Therefore you have to subtract away the overlap so those items are only counted once.

Figure 2 - taking care of the overlap

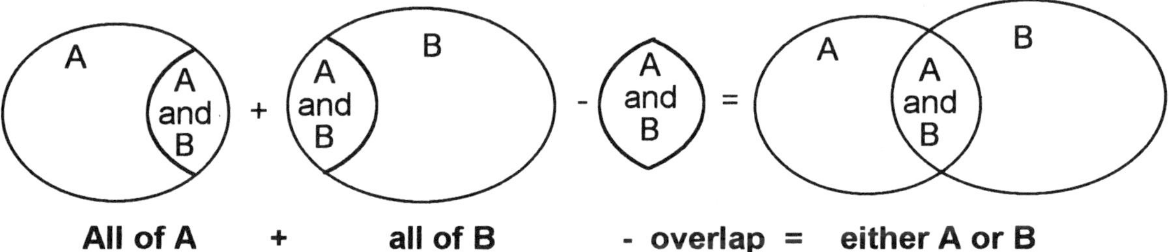

What we have called the **overlap** is the **intersection** of two sets in set theory, and is related to statements containing the word "**and**". The **"either A or B or both"** is called the **union** in set theory, and is related to statements containing the word "or".

The cars/white/cassette problem needs both rule 1 and then rule 2. With rule 1, the total number of people with cars which are white or have a cassette deck (or both) = 170 + 330 - 100 = 400. Then the number with neither can be found with rule 2: 500 - 400 = 100. Thus the number of people surveyed who had a non-white car without a cassette deck is 100.

⌘1 A survey was done on a group of cat lovers. 50 had long-haired cats, and 70 had white cats, and 23 had white long-haired cats. How many who were surveyed had white short-haired cats?

⌘2 A mountain-climbing club has 130 members. 70 of them had climbed Mt. McKinley, and 60 of them had climbed Mt. St. Helens (before the volcano in it exploded). 40 of them had climbed neither. How many of them had climbed both?

6.1 Exercises

1. Out of the 150 flowers in a garden, 55 are roses, 65 are white flowers, and 15 are white roses. How many of the flowers are neither white nor roses?
2. A store has many different kinds of socks for sale: 150 of the socks are made from cotton, 120 of them are black, and 30 socks are black socks made of cotton.. How many socks are cotton but not black?
3. A kitchen has a number of pans in it. 38 of the pans have lids and 31 have copper bottoms, and 18 have both lids and copper bottoms. How many have copper bottoms only?
4. A restaurant has three choices for the fast meal: meat and soup, meat and salad, or meat with both soup and salad. Last night the restaurant served 92 meals. If the restaurant sold 20 meat and soup meals, 68 meat and salad meals, and 15 meat with soup and salad, how many meals did it sell which were not a fast meal?
5. Out of the 400 farmers attending a convention, 150 showed up early for the pre-convention seminar, 130 stayed over for the post-convention seminars, and 40 did both. How many attended neither the pre- or post- convention seminars?
6. A restaurant serves a special on enchiladas and egg rolls on Thursdays. Of the 360 meals sold last Thursday, 200 were egg rolls only, 75 were enchiladas only, and 67 meals had both. How many meals were sold which had neither?
7. A pop music lover's club has 1500 members. of which 400 are exclusively country music fans, 150 are fanatical rap-only lovers, and 300 who like neither. How many like both kinds of music?
8. Last weekend at a quick-stop convenience store, they served 120 people. 80 of the customers bought something to drink, 60 bought something to eat, and 40 bought both. How many customers bought neither food nor drink?
9. 500 students enrolled for the junior year this year. 170 enrolled for both a math and an English course and 130 enrolled for nether. How many enrolled for either an English course or a math course, but not both?
10. A literature club has 460 members. If a survey of the members revealed that 255 of them liked both historical and romantic novels, and 50 liked neither, how many liked either type but not both?
11. A stereo outlet has 640 stereos, some of which have built-in CD players, some have built-in cassette decks, and some have both. If 370 of the stereos come with CD player, and 130 have both CD and cassette, and, 230 have neither, how many have cassette only?
12. A large ranch did inventory on its horses, and found that out of its 165 work horses, 85 were part Arabian, 65 were part Palomino, and 35 had neither Arabian nor Palomino ancestors. How many had both kinds of ancestors?

6.2 Order Problems

This kind of problem is also called a "bench problem", "seating problem", or "race problem", referring to some question about order in time or space. You are supposed to come up with an order given various clues. There is no standard way to work this kind of a problem, although making a rough drawing usually helps. The rest is up to your deductive powers. Working through some examples may show you some possible methods.

- (easy) Marie, Jaime, Susie, and George were in a foot race. Susie came in behind Jaime, and Marie came in ahead of Jaime, and George came in between Jaime and Susie. Who came in third?.

  Working it out . . .
  "Susie came in behind Jaime." (lead) Jaime Susie
  "Marie came in ahead of Jaime." (lead) Marie Jaime Susie
  "George came in between Jaime and Susie"
  (lead) Marie Jaime George Susie
  Answer: George came in third.

- (medium) The gambler Bad Lad Brown spread out his straight so that everyone at the table could see all 5 cards (4 - 8). The 8 was on one of the ends, and the 7 was separated from it by at least two cards. The 5 was to the right of the 4, and the 6 was to the left of the 4. The 5, which was not on an end, was next to the 7.

  Working it out . . .
  "The 8 was on one of the ends, and the 7 was separated from it by at least two cards."
  Possibilities : 8 – – 7 – 8 – – – 7 – 7 – – 8 7 – – –8
  "The 5 was to the right of the 4, and the 6 was to the left of the 4."
  That means: 645
  Possibilities: 86475 86457 67485 76458
  "The 5, which was not on an end, was next to the 7."
  Possibility: 86457
  The answer is 86457.

- (tough) Alfredo, Ralph, Johnny, and Manuel were spending Spring Break in a Condo on the beach. They each had a room, and two of them had access to a common balcony (see diagram in **Drawing 1**). Working with the information below, who stayed in room A?

  1. Ralph is frustrated because he does not have a room opening on the balcony.
  2. Johnny has a room with a door opening onto the balcony.
  3. Manuel has a room next to the kitchen.
  4. Alfredo absolutely insists on a room which is near both the balcony and the kitchen.

Working it out by thinking it through in successive drawings: Taking it statement by statement:

Drawing 1

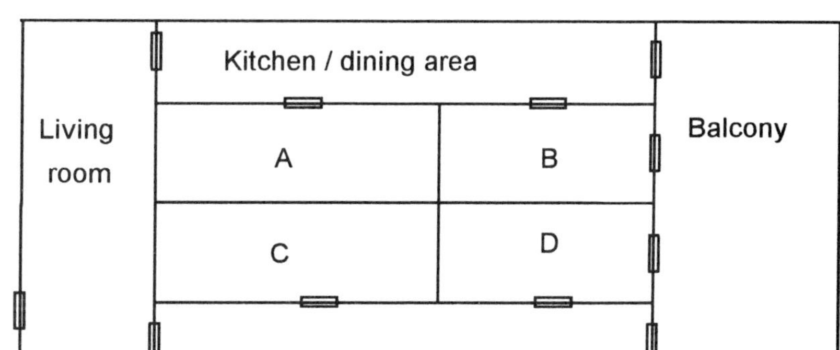

1) Ralph is frustrated because he does not have a room opening on the balcony. This means that Ralph is in A or C

Drawing 2

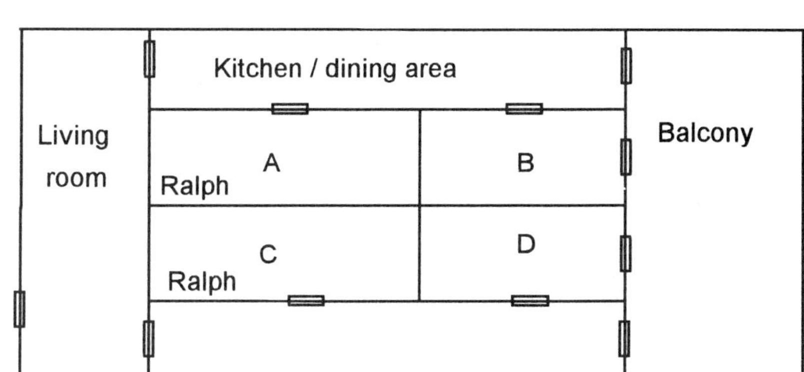

2) Johnny is in a room which has a door opening directly onto the balcony. This means that Johnny is in B or D

Drawing 3

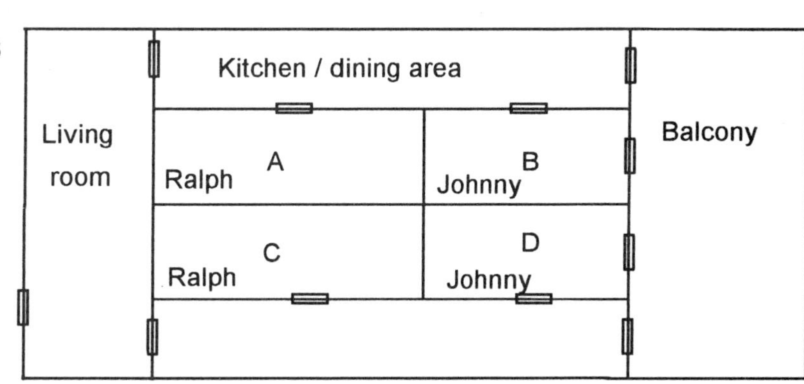

Manuel has a room next the kitchen. This means that Manuel is in either A or B.

Drawing 4

Kitchen / dining area

Living room

Manuel
A
Ralph

Manuel
B
Johnny

Balcony

Ralph
C

D
Johnny

Alfredo absolutely insists on a room which is next to both the kitchen and the balcony.

Drawing 5

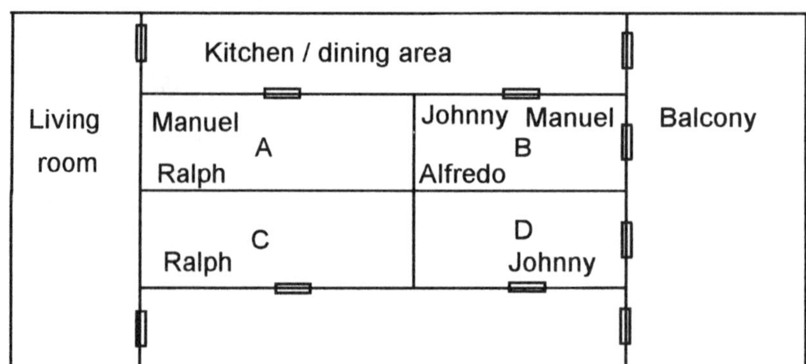

This means that Alfredo must be in room B. This forces Johnny to room D and Manuel to room A, and Ralph to room C.

Drawing 6
The solution

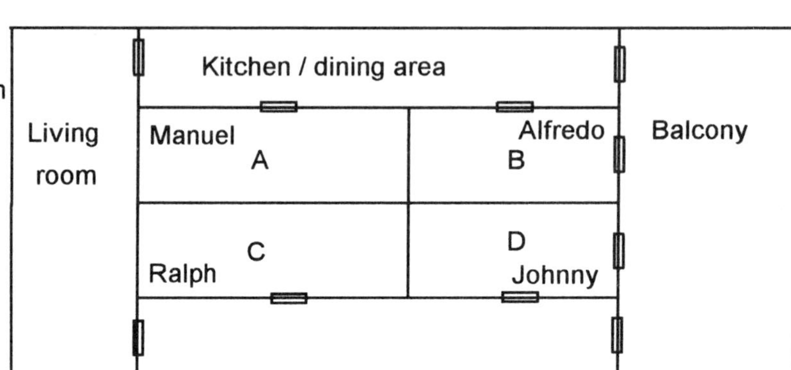

6.2 Exercises

1. Maria, Ana, and Catalina are in a foot race. Ana came in before Catalina. Catalina did not beat Maria. Maria did not win the race. Who came in first?

2. Caroline, Darlene, Esmarie, and Felice were the Ace High School entrants in the national Spelling Bee.
   Darlene did not finish last for her school.
   Esmarie beat Darlene
   Caroline's spelling rating always trails Darlene's.
   Felice never beats Esmarie
   Darlene was third.
   Who was first?

3. There is a dorm TV room with very comfortable chairs. An opaque door leads into the room. Lucinda went into the room. Marco went into the room to see if Raul was there, and left because Raul was not there. Sarita went in to talk to Lucinda if she was there, or to leave otherwise. Pepe and Sarita came out of the room looking for some popcorn, and ended up talking to Raul. After Lucinda came out, she and Raul and Pepe made some popcorn and then went back into the TV room. Sarita decided to catch the evening news, and went into the room to see it. Lucinda got tired of the guys talking, and left the room. At this point, who was in the TV room?

4. A disabled student needs help in setting up her Tuesday-Thursday class schedule. She needs to take three $1\frac{1}{2}$ hour classes on Tuesday and Thursday.

   The class times are 8 - 9:30, 9:30 - 11, 11 - 12:30, 12:30-2, 2-3:30, 3:30 - 5.
   The classes she wants to take are Swimming, Algebra, and African History.
   Her favorite History teacher teaches the 9:30 African History class.
   She needs to have her lunch between 11 and 2
   She cannot have any physical education class right after lunch, or right before or after any other class (because of the distance to the gym area).
   None of the classes she wants to take are offered at 3:30
   She likes to get her hardest class, Algebra, out of the way first
   What class schedule should she aim for?

5. The Mason County Turtle Race results are finally in:
   Speed O'Lite came in before Bongo
   Consternation moved faster than Bongo.
   Pretty in Puce always beats Consternation.
   Consternation beat Speed O'Lite
   Who came in third?

6. Alpine is 10 miles north of Bedrock and 10 miles east of Canyonville. Dust Valley's closest neighbor is 5 miles to the south. It's further to Dust Valley from Alpine than from Canyonville. All roads go either north, south, east, or west.
   How far is it from Bedrock to Dust Valley?

7. There are 4 places on a bus bench. Pete, Ralph, Scott, and Tom sit on the bench every day waiting for the bus to take them home. Ralph likes to sit on the West end of the bench and work on his suntan. Scott refuses to sit by Ralph, but Pete likes to sit by Ralph because he is such a good listener. If today Tom did not sit on the East end of the bench, who did?

8. There are four stalls in a row which hold 4 race-horses: | 1 | 2 | 3 | 4 |
   As You Wish, *Banana Split*, *Can't Beat'em*, and *Donut Lover*.
   Stall 3 is always occupied by either *As You Wish* or *Banana Split*.
   Stall 2 always has *Can't Beat'em*.
   Donut Lover bites *As You Wish* and *Banana Split* so he can not be in a stall next to either of them.
   Banana Split likes *Can't Beat'em*, so he usually gets a stall next to him.
   Who is in Stall 4?

6.3 Deductive Puzzle Problems

Deductive puzzle problems are sometimes called "grid" puzzles after the recommended method of solution. These puzzles usually start like this:

- Abe, Betty, Connie, and Doug are a lawyer, a doctor, a teacher, and a Realtor, but not necessarily in that order. . .

Then follow several statements which allow you, by process of elimination, to figure out who has which vocation. These are most easily solved using a grid - an array of rectangular Venn diagrams, and successive deductions via logical arguments, using these diagrams. Although these problems can be solved without using a grid, the grid is probably the least confusing method. In the kind of puzzles you are most likely to encounter, the statements contain no quantifiers.

This is how you would work through a grid puzzle:

- Andy, Ben, Charlie, and Doug are good friends and members of a golf foursome. They are a pharmacist, doctor, surgeon, and male nurse, but not necessarily in that order.
    1) Last week Doug helped the doctor choose a gift for his wife.
    2) Ben and the surgeon like to challenge each other at handball.
    3) Charlie, the surgeon, and the pharmacist eat lunch together as duties allow.
    4) Andy carpools with the doctor.
    5) The male nurse likes to unwind after a tough day by playing 3-man basketball with Ben and Charlie.
    6) Andy never studied surgery.

    Who is the surgeon?

To solve this kind of problem, set up a grid with the names across the top and the vocations down the side.**(3a)** (You can do it with the names down the side and the vocations across the top if you prefer.) If you are fortunate enough to have names and vocations starting with different letters, all you need is the first letter.**(3b)**

Figure 3 - setting up the grid

a)

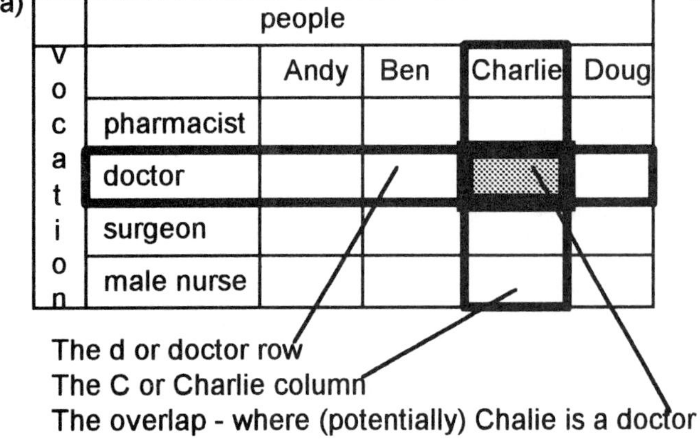

or

b)

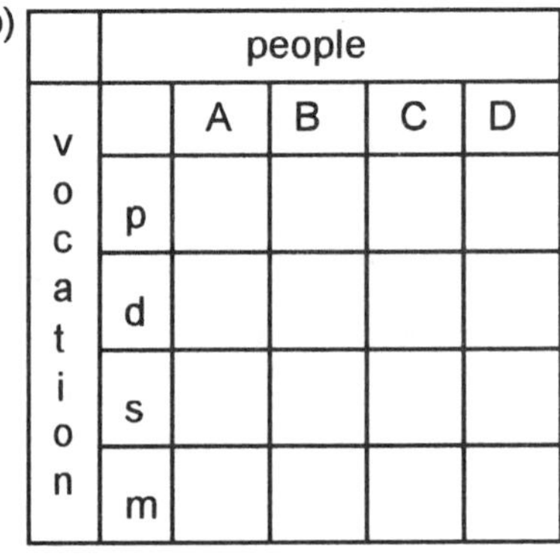

Once the grid is set up, take each statement as it comes and mark the grid. The first statement, 1) Last week Doug helped the doctor choose a gift for his wife. implies that Doug is not the doctor (he could not both be the doctor and help the doctor at the same time). We show this by putting an **X** in the box where the **D**oug column intersects the **d**octor row **(4a)**. In the same way, 2) Ben and the surgeon like to challenge each other at handball. implies that Ben cannot be the surgeon **(4b),** so the **B**en-**s**urgeon box gets an **X.** The statement 3) Charlie, the surgeon, and the pharmacist eat lunch together as duties allow. is shown in **(4c)** with an **X** in the **C**harlie-**s**urgeon and **C**harlie-**p**harmacist boxes.

Figure 4 - working out the problem with the grid

a) Doug not the doctor

vocation \ people	A	B	C	D
p				
d				**X**
s				
m				

b) Ben not the surgeon

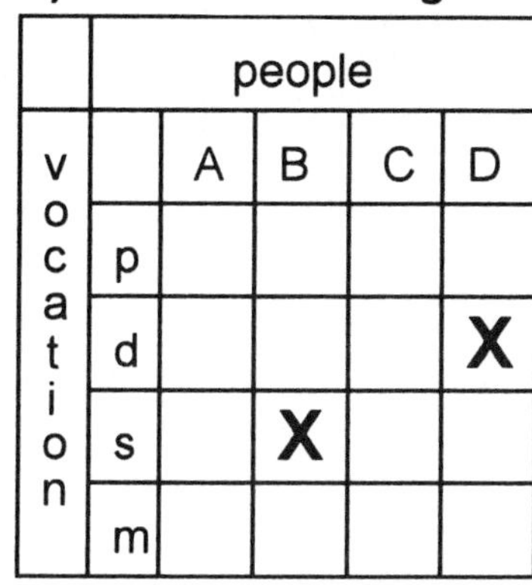

vocation \ people	A	B	C	D
p				
d				**X**
s		**X**		
m				

c) Charlie not the surgeon or the pharmacist

vocation \ people	A	B	C	D
p			**X**	
d				**X**
s		**X**	**X**	
m				

4) Andy carpools with the doctor. is shown in **(4d)** with an **X** in the **A**ndy-**d**octor box. 5) The male nurse likes to unwind after a tough day by playing 3-man basketball with Ben and Charlie. is shown by **(4e).** This statement puts **X**'s in two boxes in the **m** row (the **B** and **C** columns). This creates a column (the **C** column) which has 3 of its 4 boxes filled - this means that **C**harlie can only be a doctor. We mark this fact by putting a **gray square** in the **C-d** box.**(4f)**

d) Andy not the doctor

vocation \ people	A	B	C	D
p			**X**	
d	**X**			**X**
s		**X**	**X**	
m				

e) male nurse not Ben or Charlie

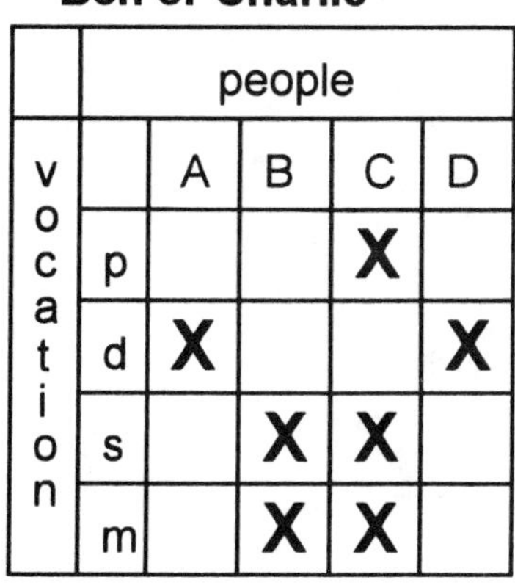

vocation \ people	A	B	C	D
p			**X**	
d	**X**			**X**
s		**X**	**X**	
m		**X**	**X**	

f) with three x's in the C column, the only choice left for C is p, so put a gray box there

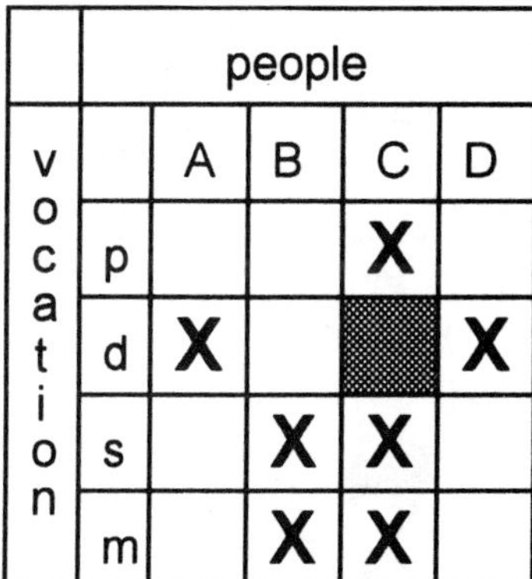

vocation \ people	A	B	C	D
p			**X**	
d	**X**		■	**X**
s		**X**	**X**	
m		**X**	**X**	

This starts a domino-type effect - now the only person who could be a **doctor** is **Charlie,** so the other boxes on row **d** get **X**'s.**(4g)** This makes the **p**-row box the only open one in the **B** column, so **Ben** has to be a **pharmacist.** Thus the **p-B** box gets a **gray square.(4h)** Then the rest of the **p**-row gets **X**'s.**(4i)**

g) X out the rest of the d row

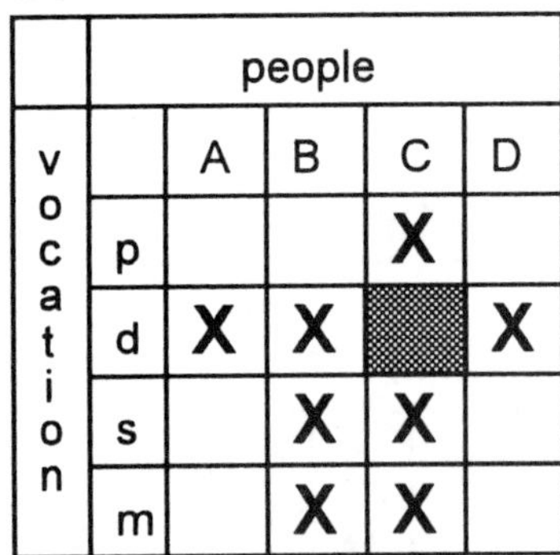

		people			
vocation		A	B	C	D
	p			X	
	d	X	X	■	X
	s		X	X	
	m		X	X	

h) the X in the d row makes the only possible choice for p to be B

		people			
vocation		A	B	C	D
	p		■	X	
	d	X	X	■	X
	s		X	X	
	m		X	X	

i) x out the rest of the p row

		people			
vocation		A	B	C	D
	p	X	■	X	X
	d	X	X	■	X
	s		X	X	
	m		X	X	

6) Andy never studied surgery. This statement puts an **X** in the **A-s** box.**(4j)** This sets off another series of domino effects. **A**ndy must be the **m**ale nurse. so a **gray square** goes in the **A-m** box.**(4k)** This means that the rest of the **m**-row gets **X**'d out.**(4l)**

j) Andy not the surgeon

		people			
vocation		A	B	C	D
	p	X	■	X	X
	d	X	X	■	X
	s	X	X	X	
	m		X	X	

k) so A is m

		people			
vocation		A	B	C	D
	p	X	■	X	X
	d	X	X	■	X
	s	X	X	X	
	m	■	X	X	

l) X out the m row

		people			
vocation		A	B	C	D
	p	X	■	X	X
	d	X	X	■	X
	s	X	X	X	
	m	■	X	X	X

That leaves only the **s** box in **D**oug' column free**,** so put a **gray square** there.**(4m)** Who is the surgeon? Doug is the answer.**(4n)**

m) this makes D be s

		people			
vocation		A	B	C	D
	p	X	■	X	X
	d	X	X	■	X
	s	X	X	X	■
	m	■	X	X	X

n) the solution - the surgeon is Doug

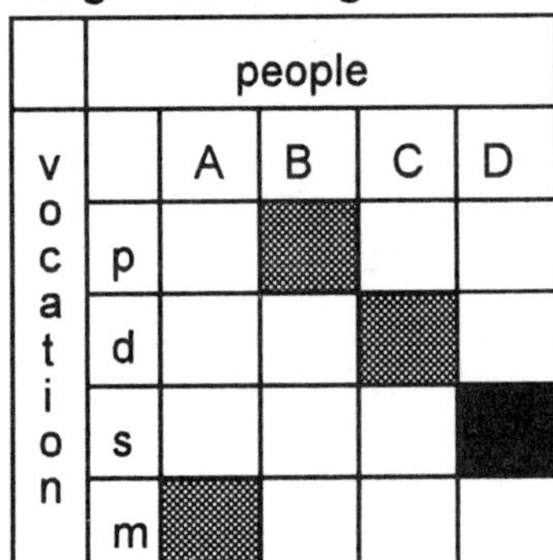

		people			
vocation		A	B	C	D
	p		■		
	d			■	
	s				■
	m	■			

o) Note that there is only one gray square per row or column.

This unabiguously says that Andy is the male nurse, Ben is the pharmacist, Charlie is the doctor, and Doug is the surgeon

The grid method shown above is the recommended method of solution. The only possible alternative way of solution - treating these clues as logic statements in an argument - requires you to keep so much going in your mind at the same time to be practical.

⌘ 3 Anna, Beatrice, Cathy, and Debby are great shopping buddies. When they are not shopping, they are an executive, a caterer, a lawyer, and a florist, not necessarily in that order.

1) Cathy likes to eat lunch with the lawyer and the executive.
2) Anna and the florist are best friends.
3) Beatrice and Debby think that the lawyer works too hard
4) Debby, the caterer, and the florist plan to shop the mega-mall in Houston next month
5) Beatrice called the caterer to make plans for an office party.

Who is the florist?

Further practice in a reasonably difficult level of this kind of puzzle is available in many monthly puzzle and crossword puzzle magazines. If you have the notion to try more difficult problems, see various advanced puzzle books (especially by Lewis Carroll, who was a pioneer in this area).

Exercises 6.3

1. Emilio, David, Gilbert, and Henry are an artist, a bookkeeper, a carpenter, and a dentist, not necessarily in that order.
   The bookkeeper will help David and Henry with their taxes.
   Henry went to the dentist today for a filling.
   Gilbert employed the carpenter and the artist redecorate his business's foyer.
   Emilio has dinner with the bookkeeper and the dentist at least once a week.
   Henry always wanted to be a carpenter, but lacked the training.
   Who is the carpenter?

2. Jose, Bert, Carlos, and Dennis are a gardener, a fireman, a policeman, and a cook, not necessarily in that order.
Jose meets twice a week with the cook and the gardener for handball.
Carlos likes to watch *Golf Today* with his Monday night golf partner, the gardener.
Bert knows he can call on the fireman and the policeman to help him move Friday.
Dennis and Bert are bridge partners with the cook.
Jose was saved once by the fireman.
Who is the fireman?

3. Andrew, Blanca, Larry, and Valerie are a karate instructor, a locksmith, a mortician, and a veterinarian, not necessarily in that order.
Andrew shares an apartment with the karate instructor and the veterinarian.
The karate instructor is a good friend of Blanca.
The mortician takes sides with Valerie against the locksmith in bridge games.
The veterinarian and Blanca are quite good friends.
The veterinarian likes to go to movies with Larry.
Andrew has an apartment next to the mortician.
What does Blanca do?

4. Adolfo, Gina, Lorraine, and Ruby are a contractor, an architect, a banker, and an engineer, not necessarily in that order.
Gina has a planning breakfast with the architect and the engineer once a month
Adolfo meets with the banker at least once a week
The engineer went to school with Adolfo and Ruby.
Adolfo's cousin is the contractor;
The banker thinks very highly of Gina's business aptitude.
Who is the contractor?

5. Bryan, Carlos, Ernest, and Lisa are a dentist, a florist, a plumber, and a therapist, but not necessarily in that order.
Bryan lives on the same block as the therapist and the plumber.
Lisa is arranging a surprise party for the dentist and the plumber next week.
The dentist treated Bryan and Ernest for plaque last week.
Who is the therapist?

6. Anna, Christina, Gerardo, and Yvette are a chauffeur, a musician, a photographer, and a travel agent, but not necessarily in that order.
The chauffeur likes to go sight-seeing on vacation tours with Christina and Gerardo
Yvette went to the travel agent yesterday to arrange a vacation trip.
Christina and the musician like to do classical works together
Anna and Christina sometimes help the photographer.
The chauffeur sometimes gets special tickets for Anna.
What is Yvette's job?

Answers and some reasons to ⌘ problems: Chapter 6

⌘ 1.

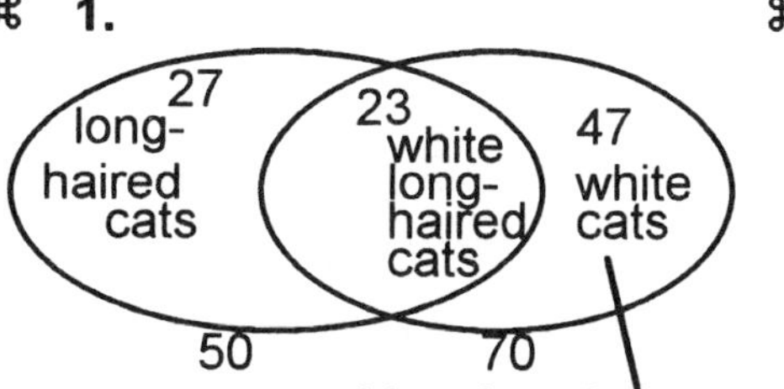

Answer: 47

⌘ 2.

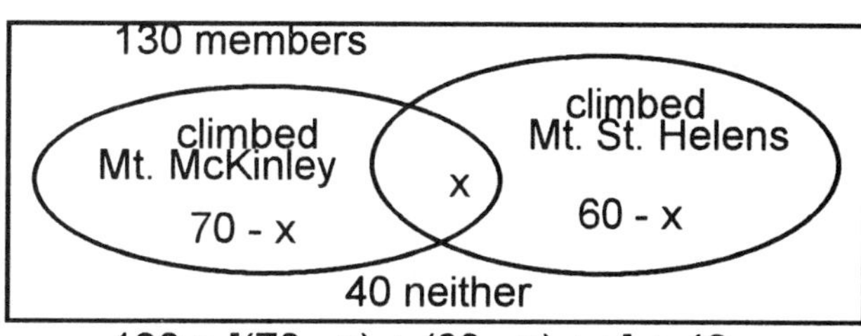

130 = [(70 - x) + (60 - x) + x] + 40

Answer: 40

⌘ 3 The grids for solving this problem go as follows:

first gray square

vocation \ people	A	B	C	D
e			x	
c				
l	■	x	x	x
f	x			

Then

vocation \ people	A	B	C	D
e	x		x	
c	x			
l	■	x	x	x
f	x			

4)

vocation \ people	A	B	C	D
e	x		x	
c	x			x
l	■	x	x	x
f	x			x

1)

vocation \ people	A	B	C	D
e			x	
c				
l			x	
f				

2)

vocation \ people	A	B	C	D
e			x	
c				
l			x	
f	x			

3)

vocation \ people	A	B	C	D
e			x	
c				
l		x	x	x
f	x			

the second gray square

vocation \ people	A	B	C	D
e	x		x	■
c	x			x
l	■	x	x	x
f	x			x

then

vocation \ people	A	B	C	D
e	x	x	x	■
c	x			x
l	■	x	x	x
f	x			x

5)

vocation \ people	A	B	C	D
e	x	x	x	■
c	x	x		x
l	■	x	x	x
f	x			x

the third gray square

vocation \ people	A	B	C	D
e	x	x	x	■
c	x	x	■	x
l	■	x	x	x
f	x			x

then

vocation \ people	A	B	C	D
e	x	x	x	■
c	x	x	■	x
l	■	x	x	x
f	x		x	x

the fourth gray square

vocation \ people	A	B	C	D
e	x	x	x	■
c	x	x	■	x
l	■	x	x	x
f	x	■	x	x

the solution:
the florist is Beatrice

ANSWERS TO ODD NUMBERED EXERCISES

EXERCISES 1.1 - 1.4

1. point : P line : m, $\overleftrightarrow{AD}$, k line segment : $\overline{BC}$, $\overline{MP}$ length : AB
   angle : $\angle ABC$, $\angle 1$, $\angle A$, $\angle M$

3. $x = 35°$ 5. $\angle 2 = 25°$ 7. $\angle 1 = \angle 5$, $\angle 2 = \angle 6$

9. $x = 75°$, $y = 50°$ $z = 55°$, $w = 55°$ 11. $x = 20°$, $y = 160°$, $z = 20°$, $w = 160°$

13. a) any two of these: $\angle 3 = \angle 6$, $\angle 4 = \angle 5$, $\angle 7 = \angle 10$, $\angle 8 = \angle 9$
    b) any two of these: $\angle 1 = \angle 5$, $\angle 3 = \angle 7$, $\angle 5 = \angle 9$, $\angle 7 = \angle 11$, $\angle 2 = \angle 6$, $\angle 4 = \angle 8$, $\angle 6 = \angle 10$, $\angle 8 = \angle 12$
    c) any two of these: $\angle 3$ and $\angle 5$, $\angle 7$ and $\angle 9$, $\angle 4$ and $\angle 6$, $\angle 8$ and $\angle 10$
    d) $\angle 7 = 118°$ e) $\angle 3 = 106°$

15. $x = 30°$ 17. $x = 30°$ 19. $x = 76°$ 21 $x = 40°$ 23. $x = 10°$

25. $x = 70°$ 27. $x = 50°$, $y = 75°$ 29. $90°$

EXERCISES 2.1 - 2.2

1. $x = 74°$ 3. $x = 50°$ 5. $x = 40°$ 7 $x = 35°$ 9. $x = 70°$

11. $x = 60°$ 13. $x = 20$ ft. 15. $x = 50°$ 17. $x = 25°$

19. If it is a right triangle then one angle is a right angle; therefore, the other two angles add up to $90°$, hence complementary

21. triangle 23. acute triangle 25. regular octagon 27. scalene

EXERCISES 2.3

1. $x = \sqrt{105}$, $x = 10.248$, $10 < x < 11$
3. $x = 5\sqrt{10}$, $x = 15.810$, $15 < x < 16$
5. $x = \sqrt{74}$, $x = 8.602$, $8 < x < 9$
7. $x = 2\sqrt{15}$, $x = 7.746$, $7 < x < 8$
9. $x = 2\sqrt{38}$, $x = 12.328$, $12 < x < 13$
11. $x = 2\sqrt{51}$, $x = 14.282$, $14 < x < 15$
13. $x = \sqrt{170}$, $x = 13.038$, $13 < x < 14$
15. $x = 2\sqrt{19}$, $x = 8.717$, $8 < x < 9$
17. $x = 3\sqrt{3}$, $x = 5.196$, $5 < x < 6$
19. $x = 4\sqrt{2}$, $x = 5.657$, $5 < x < 6$
21. $x = 10\sqrt{2}$, $x = 14.142$, $14 < x < 15$
23. $A = 240$ sq. yd.
25. $A = 360$ sq. ft.

EXERCISES 3.1 - 3.2

1. $\frac{1 \text{ in.}}{5 \text{ ft.}}$ 3. $\frac{200 \text{ g}}{1 \text{ wp}}$ 5. $\frac{5 \text{ dimes}}{3 \text{ dimes}}$ or $\frac{5 \text{ nickles}}{3 \text{ nickles}}$ 7. a) $\frac{1 \text{ taffy}}{4 \text{ assort.}}$ b) $\frac{4 \text{ bonbon}}{11 \text{ hard candy}}$

9. x = cost of shoes
x = \$45

11. x = servings
x = 37.5

13. x = profit
x = \$ 6.67

15. x = value of house
x = \$ 33,333.33

17. x = gallon
x = 420 gal.

19. x = bill for year
x = \$936

23. a) x = flour
x = 240 cups
b) x = sugar
x = 100 cups

21. x = hours
x = $11\frac{1}{4}$ hrs

25. x = feet
x = 300 ft.

27. x = gallons
x = $8\frac{3}{4} \sim 9$ gal

29. x = goose population
x = 1,666.6 ~ 1,667

EXERCISES 3.3

1. x = $16\frac{1}{3}$ in.
x = 16.33 in.

3. x = $17\frac{1}{7}$ ft.
x = 17.14 ft.

5. x = $5\frac{7}{13}$ in.
x = 5.54 in

7. x = $15\frac{15}{19}$ m
x = 15.79 m

9. x = 5 yd.

11. x = 3.6 ft.

13. a) a = $2\sqrt{5}$ in. b) b = 4 in.

15. x = 48 ft.

17. x = 9 in.

19. x = 40 ft.

21. x = 45 ft.

EXERCISES 3.4

1. RP = PN
$\angle R = \angle N$
$\angle RPQ \cong \angle MPN$
$\Delta MNP \cong \Delta QRP$
ASA ≅ ASA

3. FK = JH
$\angle K \cong \angle H$
$\angle F \cong \angle J$
$\Delta GFK \cong \Delta GJH$
ASA ≅ ASA

5. AD = DC
AB = BC
BD = BD
$\Delta DBA \cong \Delta DBC$
SSS ≅ SSS

7. TQ = PQ = SQ = RQ
$\angle TQP \cong \angle SQR$
$\Delta SQR \cong \Delta TQP$
SAS ≅ SAS

9. AD = DC
$\angle ADB \cong \angle BDC$
DB = DB
$\Delta DBA \cong \Delta DBC$
SAS ≅ SAS

11. PM = MN
LM = MO
$\angle PML \cong \angle DMN$
$\Delta PML \cong \Delta OMN$
SAS ≅ SAS
PL = ON
Corresponding parts of ≅ Δ are equal.

13. x = 60°

15. x = 20

EXERCISES 4.1 - 4.3

1. valid statement
3. not a valid statement - needs a subject
5. not a valid statement
7. not a valid statement - question
9. valid statement
11. not a valid statement - wish

(Refer to Drawing 1) Answers may vary.

13. No white cat is long-haired.
15. Some long-haired cats are happy cats.
17. Some long-haired cats are happy, and some have white fur, but no white cats are happy.
19. All happy cats have white fur, and some other white cats are long-haired.
21. All long-haired cats have white fur.

(Refer to Drawing 2) Statements may vary from person to person.

23. Duke is a dog, but he does not have long hair.
25. Midnight has long hair and is not a dog

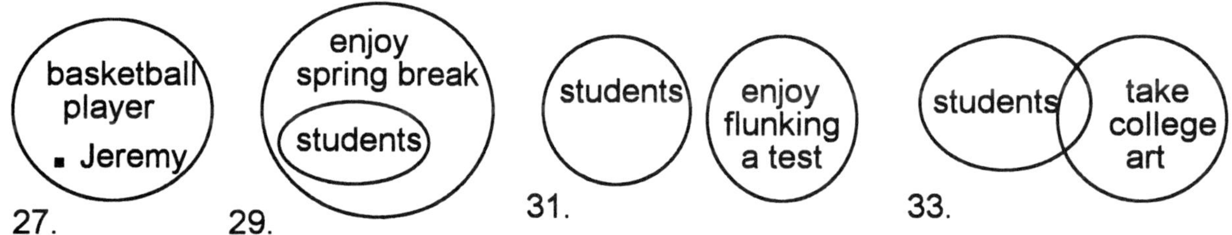

35. TRUE <u>or</u> FALSE = **TRUE**
37 TRUE <u>and</u> FALSE = **FALSE**
39. FALSE <u>or</u> TRUE = **TRUE**
41. TRUE <u>and</u> FALSE = **FALSE**

EXERCISES 4.4 - 4.5

1. Vicente is not a good athlete
3. Ellen tolerates lazy students.
5. Nelly takes tests on time.
7. All candy is made with sugar.
9. Some cooks do not have a lot of cookbooks.
11. Some hikers pass up a nice day for a hike.
13. No frozen food is low-calorie.
15. All chairs do have a back.
17. Some days do not have 24 hours.
19. Flowers do not need sun, or flowers need insect pests.
21. Angela likes geometry and Lionel likes algebra.

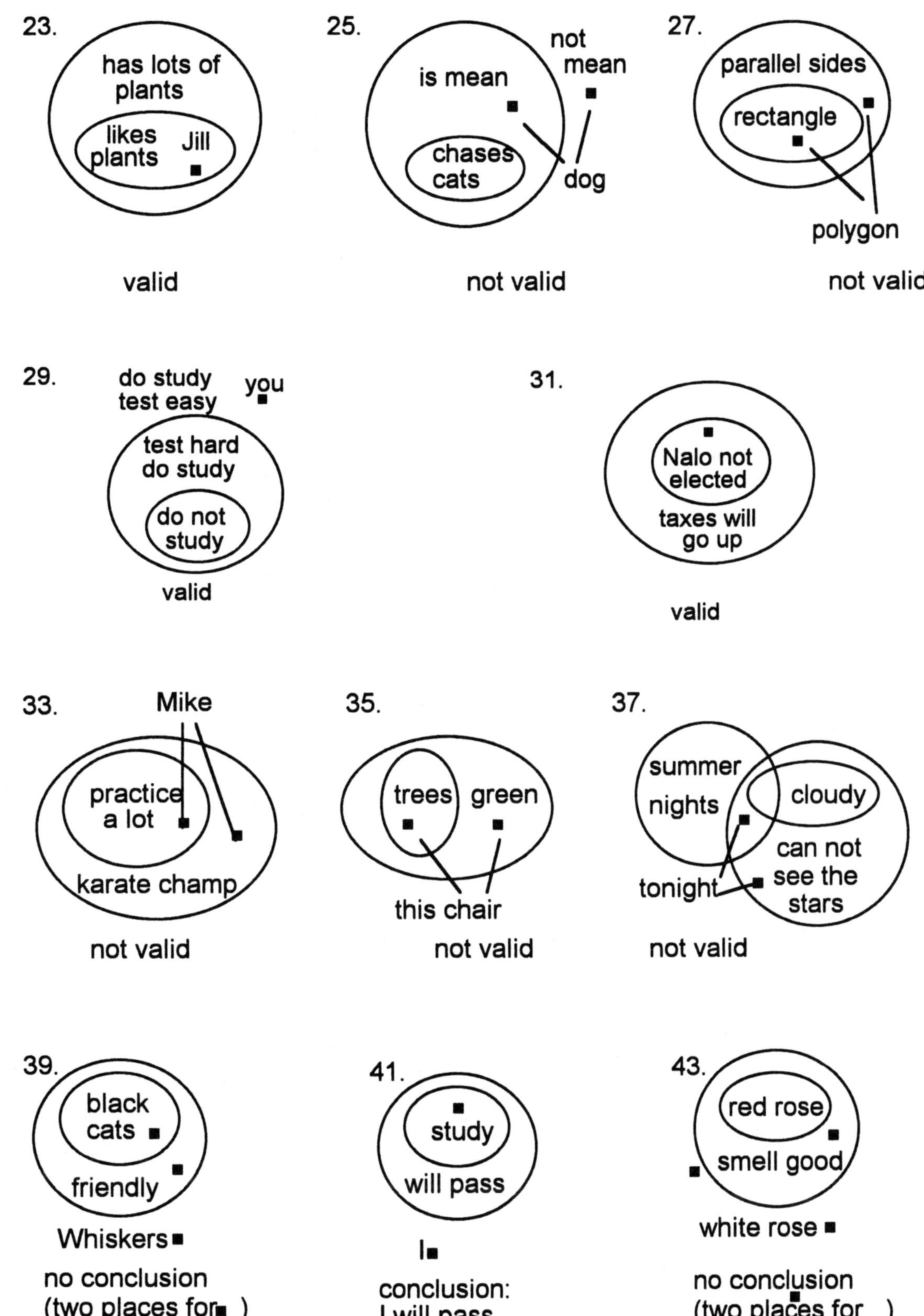

23.
has lots of plants
likes plants
Jill
valid
25.
is mean
not mean
chases cats
dog
not valid
27.
parallel sides
rectangle
polygon
not valid
29.
do study test easy
you
test hard do study
do not study
valid
31.
Nalo not elected
taxes will go up
valid
33.
Mike
practice a lot
karate champ
not valid
35.
trees
green
this chair
not valid
37.
summer nights
cloudy
can not see the stars
tonight
not valid
39.
black cats
friendly
Whiskers
no conclusion
(two places for)
41.
study
will pass
I
conclusion:
I will pass.
43.
red rose
smell good
white rose
no conclusion
(two places for)

45. statement: If Roy does not have a free weekend, he will not visit his grandkids.
converse If Roy did not visit his grandkids, he did not have a free weekend.
inverse: If Roy has a free weekend, he will visit his grandkids.
contrapositive: If Roy visits his grandkids, he had a free weekend.

47. statement: A rectangle with four equal sides is a square.
converse: If it is a square, it is a rectangle with four equal sides.
inverse: If it is not a rectangle with four equal sides, it is not a square.
contrapositive: If it is not a square, it is not a rectangle with four equal sides.

49. statement: If I have to go, then you have to go too.
converse: If you have to go too, then I have to go.
inverse: If I do not have to go, then you do not have to go too.
contrapositive: If you do not have to go too, then I do not have to go.

EXERCISES 5.1-5.3

1. 17; add 3 each time
3. 123; 2nd level diff. of 7
5. 36; 2nd level difference of 2 decreasing
7. j
9. 91; add series of square, or 3rd level difference of 2
11. 7; back and forth 2nd level difference of 2
13. P
15. 222; 3rd level difference of 6
17. 21 Fibonacci
19. $\frac{5}{6}$, denominator = numerator + 1, increasing
21. $\frac{3}{25}$, [$\frac{3}{1^2}, \frac{3}{2^2}, \frac{3}{3^2}, \frac{3}{4^2}, \frac{3}{5^2}$]
21. $\frac{1}{3}$, $\frac{2}{1}, \frac{2}{2}, \frac{2}{3}, \frac{2}{4}, \frac{2}{5}$
23. b
25. d
27. c
29. a

EXERCISES 6.1

1. 45
3. 13
5. 160
7. 650
9. 200
11. 40

EXERCISES 6.2

1. Ana
3. Raul, Sarita, Pepe
5. Speed O'Lite
7. Scott

EXERCISES 6.3

1. Emilio
3. Mortician
5. Lisa

TABLE A FORMULAS

rectangle

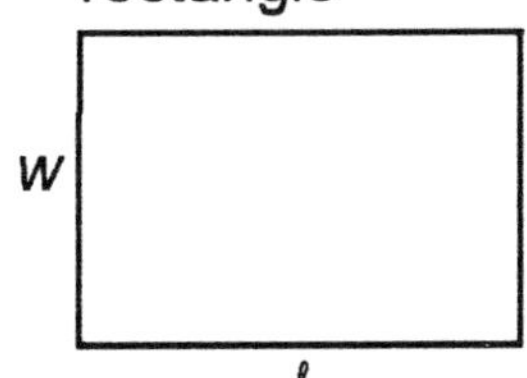

$P = 2\ell + 2w$
$A = \ell w$

triangle

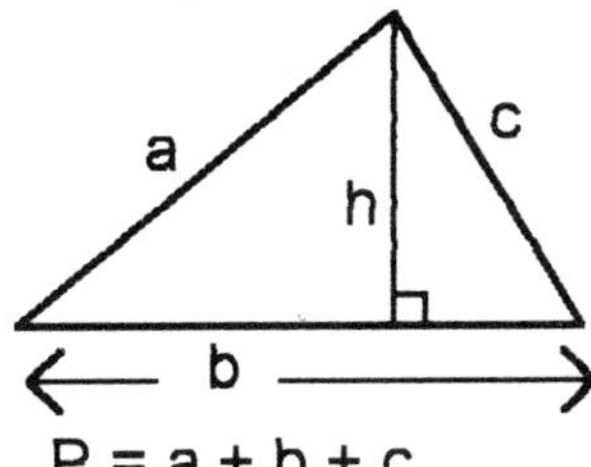

$P = a + b + c$
$A = \frac{1}{2} bh$

PYTHAGOREAN THEOREM (for right triangle)

a
c
b

$c^2 = a^2 + b^2$

square

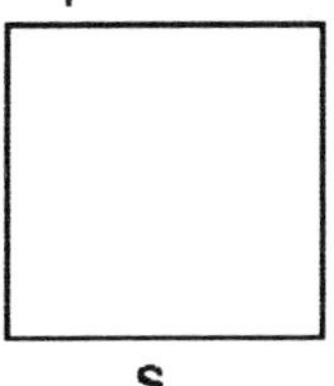

$P = 4s$
$A = s^2$

trapezoid

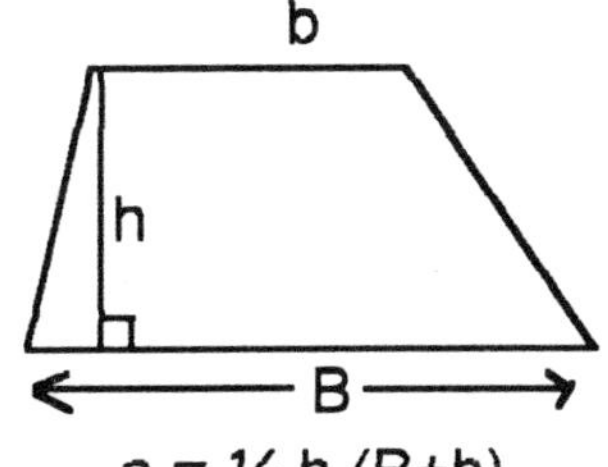

$a = \frac{1}{2} h (B+b)$

parallelogram

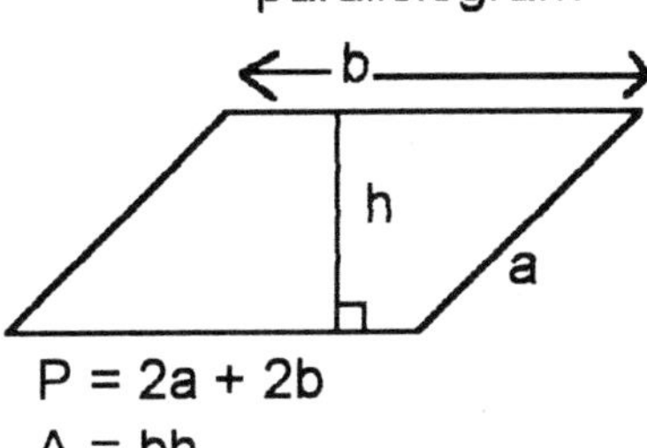

$P = 2a + 2b$
$A = bh$

circle

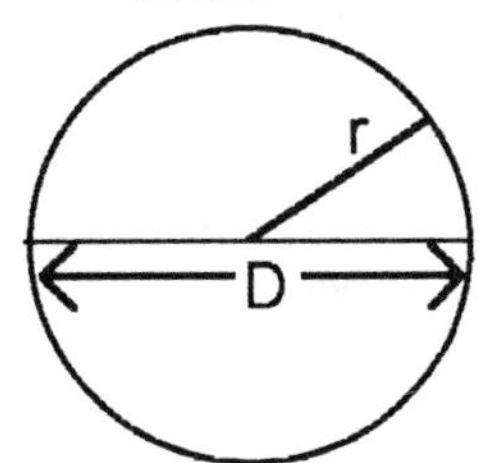

$D = 2r$
$C = 2\pi r$
$C = \pi D$
$A = \pi r^2$

rectqngular solid

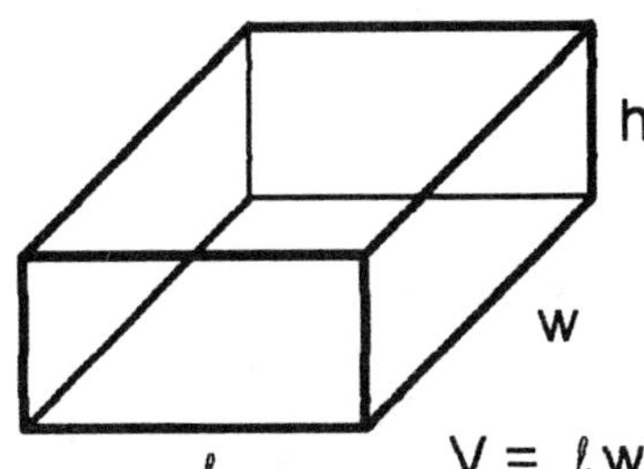

$V = \ell w h$

cube

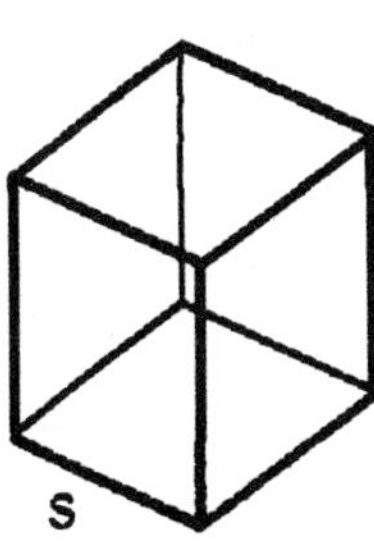

$V = s^3$

cone

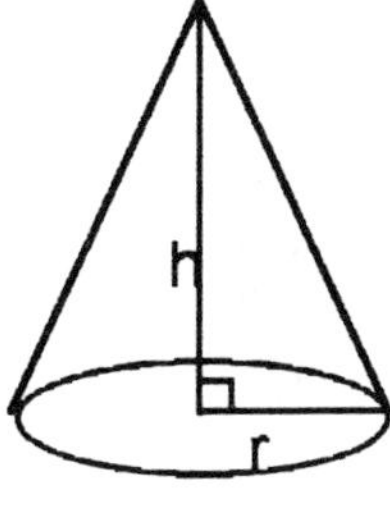

$V = \frac{1}{3} \pi r^2 h$

cylinder

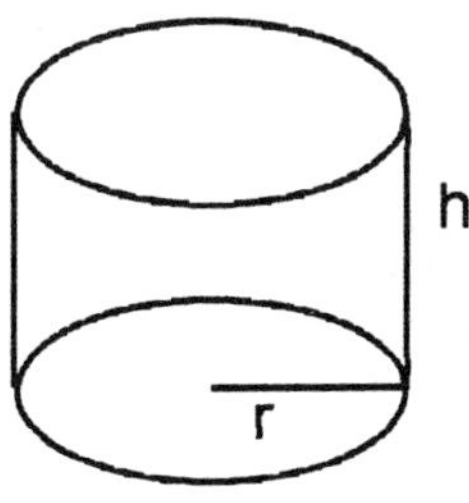

$V = \pi r^2 h$

sphere

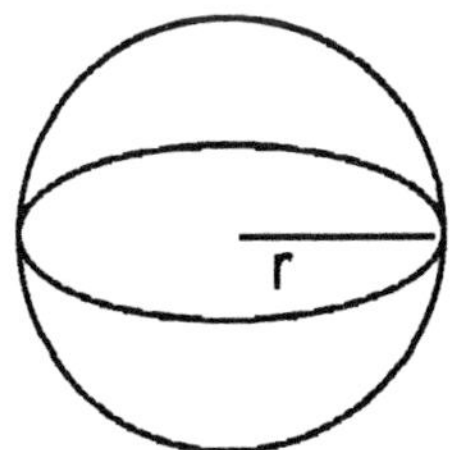

$V = \frac{4}{3} \pi r^3$

TABLE B

TABLE OF SQUARES AND SQUARE ROOTS

n	n^2	$\sqrt{n}$	n	n^2	$\sqrt{n}$	n	n^2	$\sqrt{n}$
1	1	1.000	46	2116	6.782	91	8281	9.539
2	4	1.414	47	2209	6.856	92	8464	9.592
3	9	1.732	48	2304	6.928	93	8649	9.644
4	16	2.000	49	2401	7.000	94	8836	9.695
5	25	2.236	50	2500	7.071	95	9025	9.747
6	36	2.449	51	2601	7.141	96	9216	9.798
7	49	2.646	52	2704	7.211	97	9409	9.849
8	64	2.828	53	2809	7.280	98	9604	9.899
9	81	3.000	54	2916	7.348	99	9801	9.950
10	100	3.162	55	3025	7.416	100	10000	10.000
11	121	3.317	56	3136	7.483	101	10201	10.050
12	144	3.464	57	3249	7.550	102	10404	10.100
13	169	3.606	58	3364	7.616	103	10609	10.149
14	196	3.742	59	3481	7.618	104	10816	10.198
15	225	3.873	60	3600	7.746	105	11025	10.247
16	256	4.000	61	3721	7.810	106	11236	10.300
17	289	4.123	62	3844	7.874	107	11449	10.344
18	324	4.243	63	3969	7.937	108	11664	10.392
19	361	4.359	64	4096	8.000	109	11881	10.440
20	400	4.472	65	4225	8.062	110	12100	10.488
21	441	4.583	66	4356	8.124	111	12321	10.536
22	484	4.690	67	4489	8.185	112	12544	10.583
23	529	4.796	68	4624	8.246	113	12769	10.630
24	576	4.899	69	4761	8.307	114	12996	10.677
25	625	5.000	70	4900	8.367	115	13225	10.724
26	676	5.099	71	5041	8.426	116	13456	10.770
27	729	5.196	72	5184	8.485	117	13689	10.817
28	784	5.292	73	5329	8.544	118	13924	10.863
29	841	5.385	74	5476	8.602	119	14161	10.909
30	900	5.477	75	5625	8.660	120	14400	10.954
31	961	5.568	76	5776	8.718	121	14641	11.000
32	1024	5.657	77	5929	8.775	122	14884	11.045
33	1089	5.745	78	6084	8.832	123	15129	11.091
34	1156	5.831	79	6241	8.888	124	15376	11.136
35	1225	5.916	80	6400	8.944	125	15625	11.180
36	1296	6.000	81	6561	9.000	126	15876	11.225
37	1369	6.083	82	6724	9.055	127	16129	11.270
38	1444	6.164	83	6889	9.110	128	16384	11.314
39	1521	6.245	84	7056	9.165	129	16641	11.358
40	1600	6.325	85	7225	9.220	130	16900	11.402
41	1681	6.403	86	7396	9.274	131	17161	11.446
42	1764	6.481	87	7569	9.327	132	17424	11.489
43	1849	6.557	88	7744	9.381	133	17689	11.532
44	1936	6.633	89	7921	9.434	134	17956	11.576
45	2025	6.708	90	8100	9.487	135	18225	11.619